Firefighter II
Student Applications
Essentials of Fire Fighting, 4th ed.

Developed by Susan S. Walker, John Joerschke, and Robert Hilley
Technical Review by Gordon Earhart and Michael Wieder
Edited by Barbara Adams, Carol Smith, Cindy
Pickering, and Lynne Murnane

Published by
Fire Protection Publications, Oklahoma State University, Stillwater, Oklahoma

ISBN 0-87939-156-1 (Firefighter II Student Applications, 4th ed., Essentails Curriculum)
ISBN 0-87939-152-9 (Curriculum Package for 4th ed., Essentials of Fire Fighting)

Second Edition
Fourth Printing, February 2002

Printed in the United States of America

Table of Contents

Chapter 7 — Rescue and Extrication
Lesson 7B — Vehicle Extrication & Special Rescue

Chapter 11 — Water Supply
Lesson 11 — Hydrant Flow & Operability

Chapter 12 — Fire Hose
Lesson 12 — Hose Tools & Appliances

Chapter 13 — Fire Streams
Lesson 13 — Foam Fire Streams

Chapter 14 — Fire Control
Lesson 14 — Ignitable Liquid & Flammable Gas Fire Control

Chapter 15 — Fire Detection, Alarm, and Suppression Systems
Lesson 15 — Fire Detection, Alarm, & Suppression Systems

Chapter 17 — Protecting Evidence for Fire Cause Determination
Lesson 17 — Fire Cause & Origin

Chapter 18 — Fire Department Communications
Lesson 18 — Radio Communications & Incident Reports

Chapter 19 — Fire Prevention and Public Fire Education
Lesson 19 — Pre-Incident Survey

How to Use this Book

This ***Firefighter II Student Applications*** workbook is designed to be used in conjunction with the teaching materials presented in its companion volume ***Firefighter II Instructor's Guide***. Both volumes use as their text the fourth edition of IFSTA's ***Essentials of Fire Fighting***.

The study sheets, information sheets, and chapter review tests in this workbook are intended to provide you with a means of studying the material covered in corresponding chapters of ***Essentials***. Practical activity sheets and job sheets are designed to allow you to master hands-on tasks requisite to meeting the standards for Firefighter II. Each activity sheet is keyed to the NFPA 1001 objectives(s) it addresses.

Practical activity sheets guide you in producing a *product* (e.g., knot, mathematical computation, analytical response), which is then evaluated by the instructor. Job sheets provide you with step-by-step procedures for performing a task (e.g., donning SCBA, raising a ladder, using a pitot tube). After you have practiced the task, your instructor will observe and evaluate your *performance*.

To receive the maximum learning experience from this workbook, it is recommended that you use the following procedure:

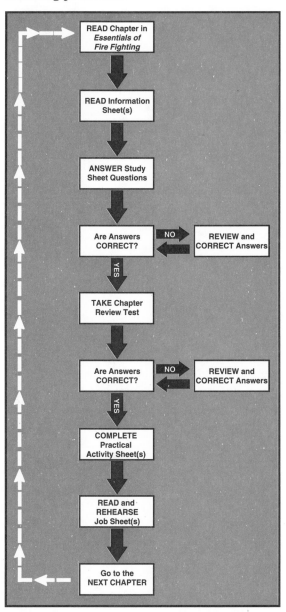

1. Before your class session date, read the assigned chapter in ***Essentials of Fire Fighting***. While reading the chapter, underline or highlight important terms, topics, and subject matter. Study the photographs and illustrations, and read the captions under each.

2. Open your ***Firefighter II Student Applications*** workbook to the corresponding lesson. Read any information sheet material included, and then follow the instructions to complete the study sheet. Answer *all* questions on the study sheet before checking your answers.

3. After you have finished, check your answers with those on the pages referenced in parentheses. (If a definition of a term is not given, or if you cannot determine the definition from the term's context, look up the term in IFSTA's ***Fire Service Orientation and Terminology*** manual or a dictionary.)

4. Correct any incorrect answers and review the material that was answered incorrectly.

5. When you have completed and checked your study sheet answers, take the chapter review test, if assigned.

6. After you have finished, check your answers against those referenced in parentheses. Jot down questions that contain content you would like to discuss or ask about during the class session.

7. Complete any assigned practical activity sheets.

8. Read any assigned job sheets. Rehearse the procedure (either mentally or physically as applicable) until you feel that you can perform the procedure competently under your instructor's observation and without the aid of the job sheet.

STUDENT APPLICATIONS

FOURTH EDITION

ESSENTIALS OF FIRE FIGHTING

LESSON I

IMPLEMENTING IMS

FIREFIGHTER II

FIRE PROTECTION PUBLICATIONS
OKLAHOMA STATE UNIVERSITY

Study Objectives

LESSON OBJECTIVE

After completing this lesson, you will be able to implement and maintain an Incident Management System and transfer command.

ENABLING OBJECTIVES

After reading Chapter 1 of *Essentials*, pages 14–20, and completing related activities, you will be able to —

1. List questions that the first person arriving at an emergency should answer.

2. **Determine the need for command.** *(Practical Activity Sheet 1-1)*

3. List the priorities of an Incident Action Plan.

4. **Organize and maintain an Incident Management System until command is transferred.** *(Practical Activity Sheet 1-2)*

5. Select facts about the transfer of command.

6. List information that should be included in a situation status report.

7. **Function within an assigned role in the Incident Management System.** *(Practical Activity Sheet 1-3)*

8. List aspects of response resources that should be tracked.

9. **Assume and transfer command within an Incident Management System.** *(Practical Activity Sheet 1-4)*

10. State the purpose of incident termination.

Study Sheet

Introduction   This study sheet is intended to help you learn the Firefighter II material in Chapter 1 of *Essentials of Fire Fighting*, Fourth Edition. You may use it for self-study, or you may use it to review material that will be covered in the lesson and chapter review tests. The numbers in parentheses are the pages in *Essentials* on which the answers or terms can be found.

Chapter Vocabulary   Be sure that you know the chapter-related meanings of the following terms and abbreviations. Use a dictionary or the glossary in *Fire Service Orientation and Terminology* if you cannot determine the meaning of the term from its context.

- Demobilization *(18)*
- IAP *(17)*
- IMS *(16)*
- Situation status report *(17)*
- Unified command *(17)*

Study Questions & Activities

1. What questions should the first person arriving at an emergency scene answer when evaluating the situation? *(16)*

    a. _____

    b. _____

    c. _____

    d. _____

    e. _____

2. List in descending order (highest to lowest) the five priorities of IAP. *(16, 17)*

    a. _____

    b. _____

    c. _____

    d. _____

    e. _____

3. How many ICs are necessary at any one time during an incident involving a single jurisdiction? *(17)*

    _____

4. What are the two main requirements for a person assuming command? *(17)*

    a. _____

    b. _____

5. Under what conditions can command be transferred via radio? *(17)*

_____

_____

6. What information should be included in a situation status report? *(17)*

a. _____

b. _____

c. _____

d. _____

e. _____

f. _____

7. How does the person assuming command acknowledge receipt of a situation status report? *(17)*

_____

8. Does command transfer automatically upon arrival of a senior officer? Explain your answer. *(18)*

_____

_____

_____

9. Who provides notification of the transfer of command: the IC relinquishing command or the IC assuming command? *(18)*

_____

10. What tracking and accountability elements should be included in an IAP? Why? *(18)*

_____

_____

_____

_____

_____

_____

11. List two actions that are accomplished as part of terminating an incident. *(18)*

a. _____

b. _____

Practical Activity Sheet 1-1
Determine the Need for Command

Name _____ Date _____

Evaluator _____ Overall Competency Rating _____

References	NFPA 1001, General 4-1.1.2 *Essentials*, pages 16–18
Prerequisites	None
Introduction	Every incident requires that someone be in charge and that an action plan be implemented. By default, the first-arriving fire department member is the Incident Commander (IC). However, as additional personnel arrive, it may be necessary to transfer command. Three questions should be asked whenever a change in command is considered:

- Does the person in command have the necessary skills and expertise to serve as IC?

- Does the person in command have the necessary level of authority to command the incident?

- Is there some other reason for transferring command (fatigue, other responsibilities, change in the situation, mutual aid response, etc.)?

Directions	Read the following scenario, and then answer the questions pertaining to it.
Activity	Off-duty firefighter Hugh Lofton notes smoke coming from the base of a chimney on the roof of a residence. He knocks on the door, but no one answers. Lofton calls 9-1-1 on his cellular telephone and reports the situation as he continues around the building. He checks the other exterior doors and finds all of them locked. As he checks the exterior of the building, he glances in windows and notes no activity in the house.

Engine Company 35 responds. Assistant Chief Marcella Chambers responds with the company, arriving 6 minutes after the call. The smoke has gotten heavier and flames are now visible around the base of the chimney.

1. When does the incident begin?

   _____

   _____

2. Who is the first IC?

   _____

   _____

3. Who is the IC when Engine 35 arrives?

_____

_____

4. Should Chief Chambers be the commander at any point in the incident? Why or why not?

_____

_____

_____

_____

5. What information should be included in the report that Lofton gives to Chambers?

_____

_____

_____

_____

_____

_____

_____

_____

6. What are some actions that Lofton might take while waiting for the fire department to respond?

_____

_____

_____

_____

_____

_____

_____

_____

_____

Competency Rating Scale

3 — **Skilled** — All 6 questions answered appropriately per suggested answers in instructor's guide and local protocol; student requires no additional practice.

2 — **Moderately skilled** — At least 4 of the 6 questions answered appropriately per suggested answers in instructor's guide and local protocol; student may benefit from additional practice.

1 — **Unskilled** — Fewer than 4 of the 6 questions answered appropriately per suggested answers in instructor's guide and local protocol; student requires additional practice and reevaluation.

☒ —**Unassigned** — Task is not required or has not been performed.

✔ **Evaluator's Note:** Score the product as indicated below. Use the rating scale above to assign an overall competency rating. Record the overall competency rating on both the student's practical activity sheet and competency profile.

To show competency in this objective, the student must achieve an overall rating of at least 2.

Answers	Correct	Incorrect
All answers evaluated per answers in **Instructor's Guide.**		
Question 1	☐	☐
Question 2	☐	☐
Question 3	☐	☐
Question 4	☐	☐
Question 5	☐	☐
Question 6	☐	☐

Practical Activity Sheet 1-2
Organize and Maintain an Incident Management System Until Command Is Transferred

Name _____ Date _____

Evaluator _____ Overall Competency Rating _____

References	NFPA 1001, General 4-1.1.2 *Essentials*, pages 16–18
Prerequisites	None
Introduction	The first-arriving department member is the first Incident Commander (IC), which means that any firefighter may have to assume command during the early phase of the response. This assignment allows you to consider some of the factors involved in incident command.
Directions	As an individual or in small groups, study the situation described below and then answer the accompanying questions/assignments.
Activity	Your engine company, along with an EMS unit, has been dispatched to a report of a vehicle accident at the side of the road. The accident has occurred within your jurisdiction on a state-maintained highway just beyond the city limits. At the scene you find that three other vehicles are now involved, including a tank truck. Two of the vehicles are actively involved in flames and at least one of them has entrapped passengers. The tank truck has gone off the road and is lying on its side. The tank is labeled "Ammonia Hydroxide" and appears to be leaking. At the base of the embankment is a canal that flows into a suburban neighborhood. The cab of the tank truck has sheared a utility pole and electrical lines are down, some of which are in contact with the truck and tank trailer. Traffic is backed up in both directions, but it is moving slowly as drivers check out the accident scene. 1. What additional resources are apparently going to be needed to assist with the response? _____ _____ _____ _____ _____

2. Considering either yourself or an officer in your department who is likely to lead an engine company on a response, would you or that person be qualified to serve as IC? Why or why not?

_____

_____

_____

_____

_____

3. List the special SOPs, mutual aid agreements, and other documentation within your department that would be required to respond to this incident. Disregard routine SOPs that are applicable to most incidents. List only those that would be required because of the unusual requirements of this incident.

_____

_____

_____

_____

4. At the bottom of the facing page, sketch an organizational chart of an appropriate response team for this incident using the resources of your department. If outside resources from other jurisdictions or other organizations are required, indicate them in your chart. Refer to *Essentials*, pages 6, 7, 14, and 18, for sample charts.

_____

_____

_____

_____

Competency Rating Scale

3 — **Skilled** — All 4 assignments completed appropriately per suggested answers in *Instructor's Guide* and local protocol; student requires no additional practice.

2 — **Moderately skilled** — At least 3 of the 4 assignments completed appropriately per suggested answers in *Instructor's Guide* and local protocol; student may benefit from additional practice.

1 — **Unskilled** — Fewer than 3 of the 4 assignments completed appropriately per suggested answers in *Instructor's Guide* and local protocol; student requires additional practice and reevaluation.

☒ —**Unassigned** — Task is not required or has not been performed.

✔ **Evaluator's Note:** Score the product as indicated below. Use the rating scale above to assign an overall competency rating. Record the overall competency rating on both the student's practical activity sheet and competency profile.

To show competency in this objective, the student must achieve an overall rating of at least 2.

Answers	Correct	Incorrect
All answers evaluated per suggested answers in **Instructor's Guide.**		
Assignment 1	☐	☐
Assignment 2	☐	☐
Assignment 3	☐	☐
Assignment 4	☐	☐

Incident Organizational Chart

Practical Activity Sheet 1-3
Function Within an Assigned Role in the Incident Management System

Name _____ Date _____

Evaluator _____ Overall Competency Rating _____

References	NFPA 1001, General 4-1.1.2 *Essentials*, pages 14–20
Prerequisites	None
Introduction	When all members of a response team understand their positions, roles, and functions in the IMS, the system can serve to safely, effectively, and efficiently use resources to accomplish the plan. In this assignment, you will explore responsibilities for those serving in various roles at an incident.
Directions	Refer to the situation described in Practical Activity Sheet 1-2. Using that situation as a representative incident, consider the roles of individuals responding to the incident. Answer the questions below with regard to the specific roles of Operations Officer, Safety Officer, and Public Information Officer in accordance with your jurisdiction's SOPs and IMS.
Activity	**OPERATIONS OFFICER** 1. Should the Operations Officer report directly to the IC or to someone else during the response? If not to the IC, to whom should the Operations Officer report? _____ _____ _____ _____ _____ 2. Should the Operations Officer be stationed at the incident site? If not, where should this person be located? _____ _____ _____ _____ _____

3. What is the primary SOP governing the Operations Officer's role at the incident?

_____

_____

_____

_____

4. What are the Operations Officer's three main responsibilities at the incident?

a. _____

_____

b. _____

_____

c. _____

_____

5. How should the operations branch be divided in terms of personnel reporting to the Operations Officer?

_____

_____

_____

_____

SAFETY OFFICER

1. Should the Safety Officer report directly to the IC or to someone else during the response? If not to the IC, to whom should this person report?

_____

_____

_____

_____

2. Should the Safety Officer be stationed at the incident site? If not, where should the Safety Officer be located?

_____

_____

_____

_____

3. What is the primary SOP governing the Safety Officer's role at the incident?

_____

_____

_____

4. What are the Safety Officer's three main responsibilities at the incident?

   a. _____ SA 1 ____

   _____

   b. _____

   _____

   c. _____

   _____

5. Does the Safety Officer have to get approval from anyone before stopping an operation at the incident site for safety reasons? If so, who must grant approval?

   _____

   _____

   _____

   _____

PUBLIC INFORMATION OFFICER

1. Should the Public Information Officer (PIO) report directly to the IC or to someone else during the response? If not to the IC, to whom should the PIO report?

   _____

   _____

2. Should the PIO be stationed at the incident site? If not, where should this person be located?

   _____

   _____

   _____

   _____

3. What is the primary SOP governing the PIO's role at the incident?

   _____

   _____

   _____

   _____

4. What are the PIO's three main responsibilities at the incident?

   a. _____

   _____

   b. _____

   _____

   c. _____

   _____

5. Does anyone within the organization have to approve the PIO's press releases before they are made public? If so, who approves them?

_____

_____

_____

_____

PAS 1-3 — Function Within an Assigned Role in the Incident Management System

---

Competency Rating Scale

3 — Skilled — All 15 questions answered appropriately per *Essentials*, pages 14-20, and local protocol; student requires no additional practice.

2 — Moderately skilled — At least 12 of the 15 questions answered appropriately per *Essentials*, pages 14-20, and local protocol; student may benefit from additional practice.

1 — Unskilled — Fewer than 12 of the 14 questions answered appropriately per *Essentials*, pages 14-20, and local protocol; student requires additional practice and reevaluation.

☒ **—Unassigned** — Task is not required or has not been performed.

✔ **Evaluator's Note:** Score the product as indicated below. Use the rating scale above to assign an overall competency rating. Record the overall competency rating on both the student's practical activity sheet and competency profile.

To show competency in this objective, the student must achieve an overall rating of at least 2.

Answers	Correct	Incorrect
*Answers based on **Essentials**, pages 14–20, and local protocol*		
OPERATIONS OFFICER		
Question 1	☐	☐
Question 2	☐	☐
Question 3	☐	☐
Question 4	☐	☐
Question 5	☐	☐
SAFETY OFFICER		
Question 1	☐	☐
Question 2	☐	☐
Question 3	☐	☐
Question 4	☐	☐
Question 5	☐	☐
PUBLIC INFORMATION OFFICER		
Question 1	☐	☐
Question 2	☐	☐
Question 3	☐	☐
Question 4	☐	☐
Question 5	☐	☐

Practical Activity Sheet 1-4
Assume and Transfer Command
Within an Incident Management System

Name _____ Date _____

Evaluator _____ Overall Competency Rating _____

References	NFPA 1001, General 4-1.1.2 *Essentials*, pages 16–18
Prerequisites	None
Introduction	The first-arriving fire department member must be prepared to transfer command to the next-arriving person with a higher level of experience or authority. A smooth and efficient transfer of command will contribute greatly to bringing the incident to a timely and successful conclusion.
Directions	With another class member, role-play transferring command for two situations. You should serve as the person assuming command in one situation and as the person relinquishing command in the other situation. Make up information as necessary, keeping the status as realistic as possible. Have paper and pencil ready, and write down all pertinent information about the transfer of command.
Activities	**TRANSFER OF COMMAND 1** Use the incident described in Practical Activity Sheet 1-1. Transfer command from Firefighter Lofton to Chief Chambers. At the time of transfer, the fire has burned through the roof of the structure at the base of the chimney. A neighbor has told Lofton that the occupants of the house are both at work for approximately three more hours. She thinks that she can find work addresses for the occupants. **TRANSFER OF COMMAND 2** Use the incident described in Practical Activity Sheet 1-2. Transfer command from one battalion chief to another because of a shift change. The fire has been extinguished in the two vehicles, and the power has been shut off in the downed electrical lines. Now the principal operations are extricating two passengers from one car, controlling the tank truck leak, and monitoring for additional fires. The state highway patrol has established traffic control. Two engine companies, two EMS units, and a haz mat unit are on site. No special equipment is needed.

Competency Rating Scale

3 — Skilled — All tasks completed appropriately per 26 criteria and local protocol; student requires no additional practice.

2 — Moderately skilled — Tasks completed appropriately per at least 10 of the 13 criteria and local protocol for each task; student may benefit from additional practice.

1 — Unskilled — Met fewer than 10 of the 13 criteria and local protocol for each task; student requires additional practice and reevaluation.

☒ **—Unassigned** — Task is not required or has not been performed.

✔ **Evaluator's Note:** Score the product as indicated below. Use the rating scale above to assign an overall competency rating. Record the overall competency rating on both the student's practical activity sheet and competency profile.

To show competency in this objective, the student must achieve an overall rating of at least 2.

Criteria	Yes	No
TRANSFER OF COMMAND 1 *(Outgoing commander)* 1. Informed relief of —		
• What has happened	☐	☐
• Whether there are injured or trapped persons	☐	☐
• What the current status is	☐	☐
• What has been done so far	☐	☐
• Whether the emergency can be handled with the resources on scene or en route	☐	☐
2. Provided complete information	☐	☐
3. Provided accurate information	☐	☐
4. Answered questions as required	☐	☐
5. Notified others of change in command	☐	☐
6. Followed department's SOPs	☐	☐
(Incoming commander) 1. Received information	☐	☐
2. Verified information by restating it to outgoing commander	☐	☐
3. Followed department's SOPs	☐	☐
TRANSFER OF COMMAND 2 *(Outgoing commander)* 1. Informed relief of —		
• What has happened	☐	☐
• Whether there are injured or trapped persons	☐	☐
• What the current status is	☐	☐
• What has been done so far	☐	☐

Criteria	Yes	No
• Whether the emergency can be handled with the resources on scene or en route	☐	☐
2. Provided complete information	☐	☐
3. Provided accurate information	☐	☐
4. Answered questions as required	☐	☐
5. Notified others of change in command	☐	☐
6. Followed department's SOPs	☐	☐
(Incoming commander)		
1. Received information	☐	☐
2. Verified information by restating it to outgoing commander	☐	☐
3. Followed department's SOPs	☐	☐

Chapter 1 Review Test

➡ **Directions:** This review test covers the Firefighter II material in Chapter 1, pages 14–20, of your ***Essentials of Fire Fighting*** text. It may be assigned as a study aid (self-test) or may be administered by your instructor as a pretest or posttest.

When used as a study aid, try to answer the questions without referring to the page numbers in ***Essentials*** or your ***Firefighter II Student Applications*** workbook *(SA)* on which the answers can be found until after you have completed the entire test. Then check your answers against those on the pages provided in parentheses.

When administered by your instructor as a pretest or posttest, read each of the test questions carefully. Choose the best response and then darken the corresponding letter on your answer sheet.

This chapter review test contains 14 multiple-choice questions, each worth 7 points. To pass the test, you must achieve at least 84 of the 98 points possible.

1. Who should initiate IMS? *(16)*

   A. The highest ranking officer within the department

   B. The highest ranking officer within the department who is on duty at the time of the incident

   C. The dispatcher

   D. The first person to arrive on the scene

2. Which of the following is ***not*** one of the first five questions that should be asked in assessing the incident upon arrival? *(16)*

   A. Does the emergency require a written IAP?

   B. What has occurred?

   C. What is the status of the emergency?

   D. Does the emergency fall within the scope of the individual's training?

3. What is the first priority of an IAP? *(16, 17)*

   A. Conducting loss control

   B. Eliminating the hazard

   C. Rescuing or evacuating endangered occupants

   D. Ensuring personnel safety and survival

4. Firefighter A says that *unified command* refers to the use of one IC when a single jurisdiction has responded.

 Firefighter B says that unified command is not an appropriate way to manage a multijurisdictional response.

 Who is right? *(17)*

 A. Firefighter A
 B. Firefighter B

 C. Both A and B
 D. Neither A nor B

5. Which of the following is **not** a correct statement about implementing an IMS? *(17)*

 A. Having sufficient resources on scene will help to ensure the safety of all involved.

 B. The organization must be structured so that all available resources can be used to achieve the goals of the IAP.

 C. To ensure unified command, the IC should avoid the use of a command staff.

 D. All incident personnel must function according to the IAP.

6. Which of the following is a correct statement about the transfer of command at an incident? *(17)*

 A. The first-arriving department member should assume command and retain it unless the incident lasts beyond a shift change.

 B. Transfer of command must be done face to face and cannot be done over the radio.

 C. Command must be assumed by someone at the scene.

 D. The transfer of command should be limited to once each 24-hour period for prolonged incidents.

7. How should the new commander acknowledge understanding the incident status upon assuming command? *(17)*

 A. By repeating the status report back to the person relinquishing command

 B. By stating, "I copy"

 C. By completing an assumption of command report

 D. By completing a situation status report

8. Which of the following is **not** required to be included in a situation status report? *(17)*

 A. A description of what happened

 B. Whether the incident fell within the expertise and authority of the person relinquishing command

 C. What resources are on the scene or en route

 D. An assessment of whether current resources are adequate

9. Transfer of command occurs ___. *(18)*

 A. When an arriving senior member decides that command should be transferred

 B. Automatically upon the arrival of a senior member

 C. Automatically when a senior member is en route to the scene

 D. Automatically after a senior member properly acknowledges receiving a situation status report

10. Firefighter A says that the IC should reduce the possibility of confusion by stating name, rank, and job title with radio transmissions.

Firefighter B says that the IC should avoid making radio transmissions because they distract firefighters from more pressing matters.

Who is right? *(18)*

A. Firefighter A   C. Both A and B

B. Firefighter B   D. Neither A nor B

11. Who should announce that command of an incident has been transferred? *(18)*

A. Incident information officer

B. Dispatcher/telecommunicator

C. Person assuming command

D. Person relinquishing command

12. How should personnel determine the size of an organizational system at an incident? *(18)*

A. Respond with all available resources and then return those that are not necessary.

B. Respond with one unit and call additional resources as required.

C. Use the smallest organization capable of handling the situation safely and efficiently.

D. Use an organization slightly larger than appears to be required so that a margin of safety is provided.

13. To track resources at an incident, the IAP should include a means of ___. *(18)*

A. Calling for additional resources, providing additional training as needed, and releasing resources when no longer needed

B. Checking in units, identifying their locations, and releasing them when no longer needed

C. Designating personnel as hot or warm zone teams and logging the duration of their exposures

D. Providing each responding unit with a radio and call sign

14. A *demobilization plan* will aid in ___. *(18)*

A. Keeping track of personnel working at the incident site

B. Recovering loaned equipment and identifying and documenting lost or damaged equipment

C. The cleanup and restoration of the site after the incident

D. Recalling resources when conditions change to make a response dangerous

REVIEW TEST ANSWER SHEET

	A	B	C	D
1.	○	○	○	○
2.	○	○	○	○
3.	○	○	○	○
4.	○	○	○	○
5.	○	○	○	○
6.	○	○	○	○
7.	○	○	○	○
8.	○	○	○	○
9.	○	○	○	○
10.	○	○	○	○
11.	○	○	○	○
12.	○	○	○	○
13.	○	○	○	○
14.	○	○	○	○
15.	○	○	○	○
16.	○	○	○	○
17.	○	○	○	○
18.	○	○	○	○
19.	○	○	○	○
20.	○	○	○	○
21.	○	○	○	○
22.	○	○	○	○
23.	○	○	○	○
24.	○	○	○	○
25.	○	○	○	○
26.	○	○	○	○
27.	○	○	○	○
28.	○	○	○	○
29.	○	○	○	○
30.	○	○	○	○
31.	○	○	○	○
32.	○	○	○	○
33.	○	○	○	○

	A	B	C	D
34.	○	○	○	○
35.	○	○	○	○
36.	○	○	○	○
37.	○	○	○	○
38.	○	○	○	○
39.	○	○	○	○
40.	○	○	○	○
41.	○	○	○	○
42.	○	○	○	○
43.	○	○	○	○
44.	○	○	○	○
45.	○	○	○	○
46.	○	○	○	○
47.	○	○	○	○
48.	○	○	○	○
49.	○	○	○	○
50.	○	○	○	○
51.	○	○	○	○
52.	○	○	○	○
53.	○	○	○	○
54.	○	○	○	○
55.	○	○	○	○
56.	○	○	○	○
57.	○	○	○	○
58.	○	○	○	○
59.	○	○	○	○
60.	○	○	○	○
61.	○	○	○	○
62.	○	○	○	○
63.	○	○	○	○
64.	○	○	○	○
65.	○	○	○	○
66.	○	○	○	○
67.	○	○	○	○

	A	B	C	D
68.	○	○	○	○
69.	○	○	○	○
70.	○	○	○	○
71.	○	○	○	○
72.	○	○	○	○
73.	○	○	○	○
74.	○	○	○	○
75.	○	○	○	○
76.	○	○	○	○
77.	○	○	○	○
78.	○	○	○	○
79.	○	○	○	○
80.	○	○	○	○
81.	○	○	○	○
82.	○	○	○	○
83.	○	○	○	○
84.	○	○	○	○
85.	○	○	○	○
86.	○	○	○	○
87.	○	○	○	○
88.	○	○	○	○
89.	○	○	○	○
90.	○	○	○	○
91.	○	○	○	○
92.	○	○	○	○
93.	○	○	○	○
94.	○	○	○	○
95.	○	○	○	○
96.	○	○	○	○
97.	○	○	○	○
98.	○	○	○	○
99.	○	○	○	○
100.	○	○	○	○

Name _____

Date _____

Score _____

Chapter 1 Competency Profile

Student Name _____ Soc. Sec. No. _____

Last   First   Middle

Fire Department _____

Address _____

Phone _____

Home Address _____

Phone _____

Date of Enrollment _____ - _____ - _____   Total Class Hours _____

Date of Withdrawal _____ - _____ - _____   Total Hours Absent_____

Date of Completion _____ - _____ - _____

Instructor's Name _____   Session Dates_____

Instructor's Directions

1. Check the candidate's competency rating (3, 2, 1, ☒) for each performance test task and psychomotor lesson objective (practical activity sheets) listed below.

2. List any additional performance tasks or psychomotor objectives (job sheets or practical activity sheets) under "Other," and check the candidate's competency rating.

3. Record the candidate's cognitive scores (written lesson tests and *administered* chapter review tests) in the spaces provided.

Level				Psychomotor Competencies
3	2	1	☒	

Practical Activity Sheets

3	2	1	☒	
☐	☐	☐	☐	PAS 1-1 — Determine the Need for Command
☐	☐	☐	☐	PAS 1-2 — Organize and Maintain an Incident Management System Until Command Is Transferred
☐	☐	☐	☐	PAS 1-3 — Function Within an Assigned Role in the Incident Management System
☐	☐	☐	☐	PAS 1-4 — Assume and Transfer Command Within an Incident Management System
☐	☐	☐	☐	Other _____
☐	☐	☐	☐	_____

Job Sheets

None Required

☐	☐	☐	☐	Other _____
☐	☐	☐	☐	_____

Level			
3	2	1	☒
☐	☐	☐	☐
☐	☐	☐	☐
☐	☐	☐	☐
☐	☐	☐	☐

Psychomotor Competencies

Chapter 1 Performance Test

Task 1 — Role-play transferring command.

Task 2 — Role-play assuming command from another firefighter.

Other _____

_____

Points Achieved	Points Needed/ Total

Cognitive Competencies

Written Test

Points Achieved	Points Needed/Total	
_____	5/5	1. List questions that the first person arriving at an emergency should answer.
		2. Evaluated on Practical Activity Sheet 1-1
_____	5/5	3. List the priorities of an Incident Action Plan.
		4. Evaluated on Practical Activity Sheet 1-2
_____	4/5	5. Select facts about the transfer of command.
_____	5/6	6. List information that should be included in a situation status report.
		7. Evaluated on Practical Activity Sheet 1-3
_____	3/3	8. List aspects of response resources that should be tracked.
		9. Evaluated on Practical Activity Sheet 1-4
_____	2/2	10. State the purpose of incident termination

Review Test

_____   Chapter 1 Review Test

Instructor's Signature _____ **Date** _____

Student's Signature _____ **Date** _____

STUDENT APPLICATIONS

FOURTH EDITION
ESSENTIALS OF FIRE FIGHTING

LESSON
3

CONSTRUCTION MATERIALS & BUILDING COLLAPSE

FIREFIGHTER II

FIRE PROTECTION PUBLICATIONS
OKLAHOMA STATE UNIVERSITY

fpp

Study Objectives

LESSON OBJECTIVE

After completing this lesson, you will be able to identify the effects of fire and fire suppression activities on structures and list actions to take when imminent building collapse is suspected.

ENABLING OBJECTIVES

After reading Chapter 3 of **Essentials**, pages 67–71, 73, and 74, and completing related activities, you will be able to —

1. Complete statements about the effects of fire and fire suppression activities on selected building materials.

2. List signs of structural instability and potential building collapse.

3. Describe ways in which fire suppression activities may create dangerous building conditions.

4. **Determine developing hazardous building or fire conditions.** *(Practical Activity Sheet 3-1)*

5. List actions to take when imminent building collapse is suspected.

Study Sheet

Introduction

This study sheet is intended to help you learn the Firefighter II material on pages 67–71, 73, and 74 of Chapter 3 of *Essentials of Fire Fighting*, Fourth Edition. You may use it for self-study, or you may use it to review material that will be covered in the lesson and chapter review tests. The numbers in parentheses are the pages in *Essentials* on which the answers or terms can be found.

Chapter Vocabulary

Be sure that you know the chapter-related meanings of the following terms and abbreviations. Use a dictionary or the glossary in *Fire Service Orientation and Terminology* if you cannot determine the meaning of the term from its context.

- Collapse zone *(74)*

- Compressive strength *(70)*

- Drywall *(67)*

- Gypsum *(67)*

- Spall *(69)*

- Tensile strength *(70)*

Study Questions & Activities

1. The reaction of wood to fire conditions depends mainly on two factors. Name them and explain why these factors are significant. *(68)*

    a. _____

    _____

    _____

    b. _____

    _____

    _____

2. Briefly describe how the following materials react to fire conditions: *(69–71)*

    a. Masonry _____

    _____

    _____

    b. Cast iron _____

    _____

    _____

    c. Steel _____

    _____

    _____

d. Reinforced concrete _____

_____

_____

e. Gypsum _____

_____

_____

f. Glass/fiberglass _____

_____

_____

3. What effect does the application of water during fire fighting have on the structural integrity of wood? *(68)*

_____

_____

_____

4. What is a common problem with masonry when water is used to extinguish chimney flue fires? *(69)*

_____

_____

5. What type of building material is rarely used in modern construction but stands up well to fire and intense heat situations? *(69)*

_____

6. List three other variables besides high temperature that can cause a specific steel member to fail. *(70)*

a. _____

_____

b. _____

_____

c. _____

_____

7. What causes gypsum to have excellent heat-resistant and fire-retardant properties? *(71)*

_____

_____

8. What is the primary objective (for the firefighter) of understanding the principles of building construction and materials? *(71)*

_____

_____

9. What are the two main causes of building collapse? *(73)*

a. _____
_____

b. _____
_____

10. Describe the differences in the likelihood of collapse between the following types of buildings: *(73)*

a. Lightweight construction *vs.* heavy timber construction

_____
_____

b. Older buildings *vs.* newer buildings

_____
_____

11. The longer a fire burns in a building, the more likely it is that the building will collapse. Why? *(73)*

_____
_____
_____
_____
_____

12. List the indicators of potential building collapse that all firefighters should be aware of and be on the lookout for at every fire. *(73, 74)*

_____
_____
_____
_____
_____
_____
_____
_____
_____
_____
_____

13. What fire fighting operations increase the risk of building collapse? *(74)*

_____
_____
_____

14. Describe the immediate safety precautions that should be taken when the collapse of a building is imminent. *(74)*

_____

_____

_____

15. How is the size of a collapse zone determined? *(74)*

_____

_____

Practical Activity Sheet 3-1
Determine Developing Hazardous Building or Fire Conditions

Name _____ Date _____

Evaluator _____ Overall Competency Rating _____

References	NFPA 1001, Fireground Operations 4-3.2a *Essentials*, pages 67–71, 73, and 74
Prerequisites	None
Introduction	Fire produces dangerous physical and chemical changes. Ventilation operations, the application of water and foam, and other fire suppression procedures also can create hazards. Firefighters must be aware of these changes and know how to respond to them. The skilled firefighter can predict likely changes and determine the best actions to take to correct the situation.
Directions	Study the illustration below. It shows a fire starting in a residence. Assume that the residence is a typical wood frame construction. A window is open to the right (south) of the fire, with a slight breeze entering the window. Evaluate how the fire will spread, the hazards that will develop as the fire spreads, and the suppression operations that should be implemented by answering the questions on the next pages.

NORTH

Activity

1. The wall immediately behind the couch on which the fire has started is a load-bearing wall in the center of the house. The wall is wood frame construction with gypsum wallboard over the studs, and wallpaper is applied to the wallboard. Describe the likely effects of the fire on this wall.

   _____

   _____

   _____

   _____

   _____

2. The ceiling is gypsum wallboard with a thin layer of textured plaster sprayed over it. The ceiling is painted. Describe the likely effects of the fire on the ceiling.

   _____

   _____

   _____

   _____

   _____

3. The walls and ceiling in the second room have similar construction and finishes as the room in which the fire began. What are the most likely routes of fire spread to the second room?

   _____

   _____

   _____

   _____

   _____

4. The department responds to the fire and opens a ventilation hole in the roof near the ridge at the north end of the house. What is the likely effect on the development of the fire?

   _____

   _____

   _____

   _____

   _____

5. An attack of the fire is made from the south window using hoselines. How is this fire suppression activity likely to affect the structure?

   _____

   _____

   _____

   _____

   _____

6. If the fire continues, how would the structure most likely collapse?

_____

_____

_____

_____

_____

Practical Evaluation

PAS 3-1 — Determine Developing Hazardous Building or Fire Conditions

Competency Rating Scale

3 — Skilled — Answers to all 6 questions are complete and appropriate per suggested answers in the instructor's materials; student requires no additional practice.

2 — Moderately skilled — Answers to at least 5 of the 6 questions are complete and appropriate per suggested answers in the instructor's materials; student may benefit from additional practice.

1 — Unskilled — Answers to fewer than 5 of the 6 questions are complete and appropriate per suggested answers in the instructor's materials

⊠ **—Unassigned** — Task is not required or has not been performed.

✔**Evaluator's Note:** Score the product as indicated below. Use the rating scale above to assign an overall competency rating. Record the overall competency rating on both the student's practical activity sheet and competency profile.

To show competency in this objective, the student must achieve an overall rating of at least 2.

Criteria	Yes	No
All responses evaluated per answers in the **_Instructor's Guide._**		
1. Described likely effects of fire on wall.	☐	☐
2. Described likely effects of fire on ceiling.	☐	☐
3. Described most likely routes of fire spreading to second room.	☐	☐
4. Described the most likely effect of north ventilation on development of fire.	☐	☐
5. Described most likely effect of fire suppression activities.	☐	☐
6. Explained how structure most likely would collapse.	☐	☐

Chapter 3 Review Test

→**Directions:** This review test covers the Firefighter II material in Chapter 3, pages 67–71, 73, and 74 of your *Essentials of Fire Fighting* text. It may be assigned as a study aid (self-test) or may be administered by your instructor as a pretest or posttest.

When used as a study aid, try to answer the questions without referring to the page numbers in *Essentials* or your *Firefighter II Student Applications* workbook *(SA)* on which the answers can be found until after you have completed the entire test. Then check your answers against those on the pages provided in parentheses.

When administered by your instructor as a pretest or posttest, read each of the test questions carefully. Choose the best response and then darken the corresponding letter on your answer sheet.

This chapter review test contains 30 multiple-choice questions, each worth 3 points. To pass the test, you must achieve at least 78 of the 90 points possible.

1. Which of the following is **not** a characteristic of composite building components such as plywood, particleboard, fiberboard, and paneling? *(68)*

   A. Highly combustible

   B. More fire resistant than solid wood

   C. Produce significant toxic gases

   D. Deteriorate rapidly under fire conditions

2. Upon what two factors does the reaction of wood to fire depend? *(68)*

   A. Size of the wood and intensity of the fire

   B. Size and moisture content of the wood

   C. Intensity and size of the fire

   D. Type of wood and size of fire

3. What may small pieces of wood be protected with to increase their resistance to heat or fire? *(68)*

   A. Masonry

   B. Steel

   C. Floated concrete

   D. Gypsum or drywall

4. Which of the following will burn the slowest? *(68)*

   A. Green wood

   B. Cured wood

   C. Glue-laminated wood

   D. Dried wood

5. What is commonly referred to as "green" wood? *(68)*

   A. Soft, flexible wood such as pine or cedar

   B. Wood with a high moisture content

   C. Wood used in new constructions

   D. Fire-retardant treated wood

6. Which of the following statements about fire retardants is correct? *(68)*

   A. They are totally effective in reducing fire spread.

   B. They are ineffective and may even be considered useless in some constructions.

   C. They are used primarily to reduce smoke and are ineffective in reducing fire spread.

   D. They are not always totally effective in reducing fire spread.

7. Firefighter A says that gypsum, like clay, becomes denser and harder under heat.

   Firefighter B says that gypsum is often used to insulate wood structural members but is of little use in insulating steel members.

   Who is right? *(71)*

   A. Firefighter A   C. Both A and B

   B. Firefighter B   D. Neither A nor B

8. What are the most commonly used materials for veneer walls? *(69)*

   A. Masonry block and brick   C. Gypsum and plaster

   B. Brick and stone   D. Cast iron and steel

9. When do masonry products most often spall and crack? *(69)*

   A. When rapidly cooled by a fire stream

   B. When exposed to high temperatures

   C. When in direct contact with fire

   D. When exposed to chemical extinguishing agents and foams

10. Near what temperature can the failure of steel structural members be anticipated? *(70)*

    A. 1,000°F *(538°C)*   C. 1,500°F *(816°C)*

    B. 1,250°F *(677°C)*   D. 1,800°F *(982°C)*

11. Which of the following statements is true? *(69)*

    A. Masonry products include bricks, cement blocks, glass blocks, and stones.

    B. Mortar deteriorates more rapidly than the masonry product it binds.

    C. Bricks are not used for load-bearing walls.

    D. Brick veneer walls are freestanding.

12. What is used as a veneer on some older buildings and can create a falling hazard during a fire? *(69, 70)*

    A. Concrete block   C. Stucco

    B. Cast iron   D. Steel

13. Firefighter A says that the temperature at which a steel member will fail depends partially on its age and its angle of inclination.

Firefighter B says that the temperature at which a steel member will fail depends partially on its composition and the load it supports.

Who is right? *(70)*

A. Firefighter A   C. Both A and B

B. Firefighter B   D. Neither A nor B

14. With what is reinforced concrete internally fortified? *(70)*

A. Gypsum-insulated I beams   C. Steel reinforcing rods or mesh

B. Cast iron reinforcement bars   D. Steel open-web trusses

15. Firefighter A says that I beams will fail in fire conditions much quicker than lightweight, open-web trusses.

Firefighter B says that elongated steel members can push out load-bearing walls and cause a collapse.

Who is right? *(70)*

A. Firefighter A   C. Both A and B

B. Firefighter B   D. Neither A nor B

16. What happens to reinforced concrete under fire conditions? *(70)*

A. It hardens and strengthens.

B. It loses strength and spalls.

C. It softens and crumbles.

D. It becomes brittle but remains structurally sound.

17. What happens when reinforced concrete is heated? *(70)*

A. Its water content is reduced, and it becomes denser and stronger.

B. Its surface becomes hot enough to ignite nearby materials.

C. Its reinforcing members warp or melt.

D. The bond between the concrete and reinforcing members might fail.

18. Firefighter A says that the length of time a steel member has been exposed to heat indicates when it may fail.

Firefighter B says that water can cool steel structural members and reduce the chance of failure.

Who is right? *(70)*

A. Firefighter A   C. Both A and B

B. Firefighter B   D. Neither A nor B

19. How many inches *(millimeters)* may a 50-foot *(15 m)* steel beam elongate when heated from room temperature to about 1,000°F *(538°C)*? *(70)*

A. 4 *(100 mm)*   C. 8 *(200 mm)*

B. 6 *(150 mm)*   D. 10 *(250 mm)*

20. What sign tells the firefighter that reinforced concrete has been damaged? *(70, 71)*

   A. Discoloration

   B. Swelling

   C. Shrinking

   D. Spalling

21. Firefighter A says that wire-reinforced glass is an effective barrier to fire extension. Firefighter B says that fiberglass insulation is noncombustible.

   Who is right? *(71)*

   A. Firefighter A

   B. Firefighter B

   C. Both A and B

   D. Neither A nor B

22. What type of product is gypsum? *(71)*

   A. Synthetic

   B. Inorganic

   C. Polymeric

   D. Organic

23. Firefighter A says that it is the sole responsibility of the safety officer to constantly monitor for unsafe building conditions.

   Firefighter B says that all structural problems should be reported to incident command personnel as quickly as possible.

   Who is right? *(71)*

   A. Firefighter A

   B. Firefighter B

   C. Both A and B

   D. Neither A nor B

24. What construction material is made of gypsum? *(71)*

   A. Plasterboard

   B. Beaverboard

   C. Fiberfill insulation

   D. Exterior sheathing

25. Why does gypsum resist heat so well? *(71)*

   A. Like brick, heat only cures the material more.

   B. It has a high intrinsic water content.

   C. Its molecules retain their excellent bonding capabilities to temperatures near 2,000°F *(1 093 °C)*.

   D. It is an excellent heat conductor and can quickly dissipate heat up to 2,000°F *(1 093 °C)*.

26. Which of the following is **not** an indicator of potential building collapse? *(73, 74)*

   A. Unusual creaks and cracking noises

   B. Unusual number of broken windows

   C. Deteriorated mortar

   D. Cracks or separations in walls, floors, and ceilings

27. The application of water to wooden structural members during fire fighting operations will **not** have a substantial effect on the ___ of the wood. *(68)*

   A. Char rate

   B. Ignition speed

   C. Strength

   D. Structural integrity

28. Firefighter A says that reinforced concrete has the compressive strength of concrete.

    Firefighter B says that reinforced concrete has the tensile strength of steel

    Who is right? *(70)*

    A. Firefighter A

    B. Firefighter B

    C. Both A and B

    D. Neither A nor B

29. How far out from a building should a collapse zone extend? *(74)*

    A. At least 300 feet *(90 m)*

    B. One and a half times the height of the building

    C. Twice as far as the height of the building

    D. A minimum of 100 feet *(30 m)*

30. Firefighter A says that dangerous building conditions may be created by firefighters trying to extinguish the fire.

    Firefighter B says that actions taken by the fire fighting team control or extinguish the fire and thus can only make the situation better.

    Who is right? *(71)*

    A. Firefighter A

    B. Firefighter B

    C. Both A and B

    D. Neither A nor B

REVIEW TEST ANSWER SHEET

	A	B	C	D
1.	○	○	○	○
2.	○	○	○	○
3.	○	○	○	○
4.	○	○	○	○
5.	○	○	○	○
6.	○	○	○	○
7.	○	○	○	○
8.	○	○	○	○
9.	○	○	○	○
10.	○	○	○	○
11.	○	○	○	○
12.	○	○	○	○
13.	○	○	○	○
14.	○	○	○	○
15.	○	○	○	○
16.	○	○	○	○
17.	○	○	○	○
18.	○	○	○	○
19.	○	○	○	○
20.	○	○	○	○
21.	○	○	○	○
22.	○	○	○	○
23.	○	○	○	○
24.	○	○	○	○
25.	○	○	○	○
26.	○	○	○	○
27.	○	○	○	○
28.	○	○	○	○
29.	○	○	○	○
30.	○	○	○	○
31.	○	○	○	○
32.	○	○	○	○
33.	○	○	○	○

	A	B	C	D
34.	○	○	○	○
35.	○	○	○	○
36.	○	○	○	○
37.	○	○	○	○
38.	○	○	○	○
39.	○	○	○	○
40.	○	○	○	○
41.	○	○	○	○
42.	○	○	○	○
43.	○	○	○	○
44.	○	○	○	○
45.	○	○	○	○
46.	○	○	○	○
47.	○	○	○	○
48.	○	○	○	○
49.	○	○	○	○
50.	○	○	○	○
51.	○	○	○	○
52.	○	○	○	○
53.	○	○	○	○
54.	○	○	○	○
55.	○	○	○	○
56.	○	○	○	○
57.	○	○	○	○
58.	○	○	○	○
59.	○	○	○	○
60.	○	○	○	○
61.	○	○	○	○
62.	○	○	○	○
63.	○	○	○	○
64.	○	○	○	○
65.	○	○	○	○
66.	○	○	○	○
67.	○	○	○	○

	A	B	C	D
68.	○	○	○	○
69.	○	○	○	○
70.	○	○	○	○
71.	○	○	○	○
72.	○	○	○	○
73.	○	○	○	○
74.	○	○	○	○
75.	○	○	○	○
76.	○	○	○	○
77.	○	○	○	○
78.	○	○	○	○
79.	○	○	○	○
80.	○	○	○	○
81.	○	○	○	○
82.	○	○	○	○
83.	○	○	○	○
84.	○	○	○	○
85.	○	○	○	○
86.	○	○	○	○
87.	○	○	○	○
88.	○	○	○	○
89.	○	○	○	○
90.	○	○	○	○
91.	○	○	○	○
92.	○	○	○	○
93.	○	○	○	○
94.	○	○	○	○
95.	○	○	○	○
96.	○	○	○	○
97.	○	○	○	○
98.	○	○	○	○
99.	○	○	○	○
100.	○	○	○	○

Name _____

Date _____

Score _____

Chapter 3 Competency Profile

Student Name _____ Soc. Sec. No. _____
            Last   First   Middle

Fire Department _____

Address _____

Phone _____

Home Address _____

Phone _____

Date of Enrollment ____ - ____ - ____   Total Class Hours _____

Date of Withdrawal ____ - ____ - ____   Total Hours Absent_____

Date of Completion ____ - ____ - ____

Instructor's Name _____   Session Dates_____

Instructor's Directions

1. Check the candidate's competency rating (3, 2, 1, ☒) for each performance test task and psychomotor lesson objective (practical activity and job sheets) listed below.

2. List any additional performance tasks or psychomotor objectives (job sheets or practical activity sheets) under "Other," and check the candidate's competency rating.

3. Record the candidate's cognitive scores (written lesson tests and *administered* chapter review tests) in the spaces provided.

Level				Psychomotor Competencies
3	2	1	☒	

Practical Activity Sheets

☐ ☐ ☐ ☐ PAS 3-1 — Determine Developing Hazardous Building or Fire Conditions

☐ ☐ ☐ ☐ Other _____

☐ ☐ ☐ ☐ _____

Job Sheets

None Required

☐ ☐ ☐ ☐ Other _____

☐ ☐ ☐ ☐ _____

Chapter 3 Performance Test

☐ ☐ ☐ ☐ Task 1 — Determine actions to take when leading an interior attack line team that encounters imminent building

Level			
3	2	1	☒

Psychomotor Competencies

collapse.

Task 2 — Evaluate and forecast a fire's growth and development.

Other _____

_____

Points Achieved	Points Needed/ Total	Cognitive Competencies

Written Test

	21/23	1. Complete statements about the effects of fire and fire suppression activities on selected building materials.
	6/7	2. List signs of structural instability and potential building collapse.
	4/4	3. Describe ways in which fire suppression activities may create dangerous building conditions.
		4. Evaluated in Practical Activity Sheet 3-1
	8/8	5. List actions to take when imminent building collapse is suspected.

Review Test

		Chapter 3 Review Test

Instructor's Signature _____ **Date** _____

Student's Signature _____ **Date** _____

STUDENT APPLICATIONS

IV

FOURTH EDITION

ESSENTIALS OF FIRE FIGHTING

LESSON 7A

RESCUE & EXTRICATION TOOLS

FIREFIGHTER II

FIRE PROTECTION PUBLICATIONS
OKLAHOMA STATE UNIVERSITY

Study Objectives

LESSON OBJECTIVE

After completing this lesson, you will be able to identify and safely use various rescue and extrication tools.

ENABLING OBJECTIVES

After reading Chapter 7 of *Essentials*, pages 186–197, and completing related activities, you will be able to —

1. Match facts about power plants to the equipment to which they apply.

2. List the two types of lighting commonly used to support emergency operations.

3. Complete statements regarding the care and use of auxiliary electrical equipment.

4. Describe guidelines for maintaining power plants and lighting equipment.

5. **Safely set up fire service lighting equipment. *(Job Sheet 7A-1)***

6. **Service and maintain portable power plants and lighting equipment. *(Job Sheet 7A-2)***

7. Identify rescue and extrication tools and equipment.

8. Match hydraulic extrication and rescue tools to their purposes.

9. List hydraulic tool safety guidelines.

10. **Use hydraulic rescue and extrication tools. *(Job Sheets 7A-3 – 7A-6)***

11. Match manual jacks and cribbing to their purposes.

12. List jacking and cribbing safety guidelines.

13. **Use manual jacks and cribbing. *(Job Sheets 7A-7 – 7A-9)***

14. Match pneumatic rescue and extrication tools to their purposes.

15. List pneumatic tool safety guidelines.

16. **Use a pneumatic chisel/hammer. *(Job Sheet 7A-10)***

17. List winch safety guidelines.

18. **Use a truck-mounted winch. *(Job Sheet 7A-11)***

19. **Use a come-along. *(Job Sheet 7A-12)***

20. Complete air lifting bag safety guidelines.

21. **Use air lifting bag(s).** *(Job Sheet 7A-13)*

22. Label the parts of a block and tackle.

23. List block and tackle safety guidelines.

24. **Use a block and tackle.** *(Job Sheet 7A-14)*

25. **Use various power saws.** *(Job Sheets 7A-15 – 7A-17)*

26. **Select correct tools for specific situations.** *(Practical Activity Sheet 7A-1)*

Study Sheet

Introduction This study sheet is intended to help you learn the Firefighter II material in Chapter 7 of *Essentials of Fire Fighting*, Fourth Edition, pages 186–197. You may use it for self-study, or you may use it to review material that will be covered in the lesson and chapter review tests. The numbers in parentheses are the pages in *Essentials* on which the answers or terms can be found.

Chapter Vocabulary Be sure you know the chapter-related meanings of the following terms and abbreviations. Use a dictionary or the glossary in *Fire Service Orientation and Terminology* if you cannot determine the meaning of the term from its context:

- Block and tackle *(196)*
- Come-along *(194)*
- Cribbing *(192)*
- Generator *(186)*
- Hydraulic *(188)*
- Inverter *(186)*

- Junction box *(188)*
- Pneumatic *(192)*
- Proof coil chain *(195)*
- Tripod *(193)*
- Twist-lock receptacle *(187)*
- Winch *(194)*

Study Questions & Activities

1. What are the major distinguishing characteristics of inverters, portable generators, and vehicle-mounted generators? *(186)*

    a. Inverters _____

    _____

    _____

    b. Portable generators _____

    _____

    _____

    c. Vehicle-mounted generators_____

    _____

    _____

2. What are the advantages and disadvantages of inverters, portable generators, and vehicle-mounted generators as compared to each other? *(186)*

    a. Inverters _____

    _____

    _____

    b. Portable generators _____

    _____

    _____

c. Vehicle-mounted generators _____

_____

3. What is the main difference between fixed and portable lighting equipment? *(187)*

a. Fixed_____

_____

b. Portable _____

_____

4. What is the purpose of a twist-lock receptacle? *(187)*

_____

_____

5. List five guidelines for servicing and maintaining power plants and lighting equipment. *(188)*

a. _____

_____

b. _____

_____

c. _____

_____

d. _____

_____

e. _____

_____

6. Name the four basic types of powered hydraulic tools used in the rescue service. *(189)*

a. *Spreaders*

b. *Shears*

c. *Combination of A + b*

d. *extension rans*

7. What is the difference between powered hydraulic tools and manual hydraulic tools? *(188)*

_____

_____

8. What was the first powered hydraulic tool to become available to the rescue service? Briefly describe how it is used. *(189)*

_____

_____

_____

_____

9. List the two most frequently used manual hydraulic tools. *(190)*

    a. _____

    b. _____

10. List both the primary advantage and the primary disadvantage of the Porta-power® system. *(190)*

    a. Advantage _____

    _____

    b. Disadvantage _____

    _____

11. What factors must be considered when using any type of jack? *(190)*

    _____

    _____

12. What is the most common use of cribbing? *(192)*

    _____

    _____

13. Cribbing can be stored in several ways. How does your department store cribbing and why was that method chosen? *(Local protocol)*

    _____

    _____

14. Why is compressed oxygen ***not*** used to power pneumatic tools? *(190)*

    _____

    _____

15. Why should caution be exercised when using pneumatic chisels in flammable atmospheres? *(193)*

    _____

    _____

16. What defines the radius of the danger zone for winch operations? *(194)*

    _____

    _____

17. What is a come-along and how is it used? *(194)*

    _____

    _____

18. What type of chain should be used for rescue work? *(194)*

    _____

    _____

19. What is the main advantage of air lifting bags? *(195)*

_____

_____

20. Name the three basic types of air lifting bags. *(195)*

    a. _____

    b. _____

    c. _____

21. What primary advantage do low- and medium-pressure bags have over high-pressure bags? *(195)*

_____

_____

22. List six safety rules that operators should follow when using air lifting bags. *(195, 196)*

    a. _____

    b. _____

    c. _____

    d. _____

    e. _____

    f. _____

23. Why is a block and tackle used for lifting heavy loads? *(196)*

_____

_____

_____

_____

24. List the basic purposes of the following tools. The first one has been done for you as an example. When you are finished, check your answers against those on the last page of the study sheet. *(186–197)*

Example: *Bar screw jack — Support*

    a. Trench screw jack _____

    b. Ratchet-lever jack _____

    c. Come-along _____

    d. Block and tackle _____

    e. Truck-mounted winch _____

    f. Hydraulic jack _____

    g. Hydraulic spreader _____

    h. Hydraulic shears _____

    i. Hydraulic extension ram _____

j. Pneumatic chisel/hammer_____

k. Pneumatic nailer _____

l. Air lifting bags _____

m. Cribbing_____

n. Tripod_____

Answers to Study Sheet question 24:

a. *Trench screw jack* — compression

b. *Ratchet-lever jack* — medium-duty lifting

c. *Come-along* — lifting, pulling

d. *Block and tackle* — lifting and pulling heavy loads

e. *Truck-mounted winch* — pulling

f. *Hydraulic jack* — heavy-duty lifting, compression

g. *Hydraulic spreader* — separation, compression

h. *Hydraulic shears* — metal cutting, spreading, pulling

i. *Hydraulic extension ram* — separation, compression

j. *Pneumatic chisel/hammer* — puncturing, cutting, driving

k. *Pneumatic nailer* — nailing wood and masonry

l. *Air lifting bags* — lifting, displacing

m. *Cribbing* — support

n. *Tripod* — providing a point for lifting

Practical Activity Sheet 7A-1
Select Correct Tools for
Specific Situations

Name _____ **Date** _____

Evaluator _____ **Overall Competency Rating** _____

References	NFPA 1001, Rescue Operations 4-4.1 *Essentials,* pages 186–197
Prerequisites	None
Introduction	As a member of a rescue team, you must be able to identify tools and select the correct tool to be used in specific situations. By mastering these skills, you may be able to save valuable time at an incident requiring rescue and quick response.
Directions	Read each situation described below and on the following page. Select the best tool for the situation from the choices provided. Write the letter of the correct choice on the blank.
Activity	_____ 1. A vehicle is overturned at the side of the road on uneven ground. The rescue team wants to stabilize the vehicle in order to continue rescue operations. In order to raise the vehicle for stabilization, which tool should be used? A. Bar screw jack B. Ratchet-lever jack C. Air lifting bags D. Cribbing _____ 2. In a building collapse, a portion of concrete slab must be raised to gain access to a void containing victims. Because of the debris, the only direction in which the slab can be moved is up. Which of the following tool(s) should be used to perform this operation? A. Winch with a tripod B. Block and tackle C. Trench screw jack D. Hydraulic jack

3. A vehicle slid off the road and is perched on the edge of a ledge. The driver is unconscious and has a head injury. The vehicle must be stabilized to prevent it from going over the edge before the victim can be removed. What approach should be used to accomplish vehicle stabilization?

    A. Chocks should be placed under the wheels on the downhill side of the vehicle.

    B. A winch should be attached to the vehicle, and the vehicle should be pulled back from the edge prior to attempting rescue.

    C. Air lifting bags should be placed under all accessible areas of the vehicle and inflated to the point that the vehicle is level.

    D. A winch should be attached to the vehicle to hold it in place, but the vehicle should not be moved while the rescue is being performed.

Practical Evaluation

PAS 7A-1 — Select Correct Tools for Secific Situations

Competency Rating Scale

3 — Skilled — All 3 questions answered correctly per answers in instructor's materials; student requires no additional practice.

2 — Moderately skilled — At least 2 of the 3 questions answered correctly per answers in instructor's materials; student may benefit from additional practice.

1 — Unskilled — Fewer than 2 of the 3 questions answered correctly per answers in instructor's materials; student requires additional practice and reevaluation.

⊠ **— Unassigned** — Task is not required or has not been performed.

✔ **Evaluator's Note:** Score the product as indicated below. Use the rating scale above to assign an overall competency level. Record the overall competency rating on both the student's practical activity sheet and competency profile.

   To show competency in this objective, the student must achieve an overall rating of at least 2.

Criteria	Yes	No
All answers evaluated per answers in **Instructor's Guide**.		
1. Selected correct tool	☐	☐
2. Selected correct tool	☐	☐
3. Selected correct tool	☐	☐

Job Sheet 7A-1
Safely Set Up Fire Service Lighting Equipment

Name _____ **Date** _____

Evaluator _____ **Overall Competency Rating** _____

References	NFPA 1001, Prevention, Preparedness, and Maintenance 4-5.2 ***Essentials,*** pages 186–188
Prerequisites	None
Student's Instructions	To meet evaluation standards, you must perform this job within _____ *[amount of time, if applicable]*; you may have _____ attempts. When you are ready to perform this job, ask your instructor to observe the procedure and complete this form. To show mastery of this job, you must perform all steps to receive an overall competency rating of at least 2.

> **Competency Rating Scale**
>
> **3 — Skilled** — Meets all evaluation criteria and standards; performs task independently on first attempt; requires no additional practice or training.
>
> **2 — Moderately skilled** — Meets all evaluation criteria and standards; performs task independently; additional practice is recommended.
>
> **1 — Unskilled** — Is unable to perform the task; additional training required.
>
> ⊠ —**Unassigned** — Job sheet task is not required or has not been performed.
>
> ✔ **Evaluator's Note:** Formulate and inform the candidate of the standards for this task (time allowed and number of attempts). Observe the candidate perform the task, check the step/key point under the appropriate attempt number as accomplished, record total time (if appropriate), and then use the rating scale above to assign an overall competency rating. If the candidate is unable to perform any step of this job, have the candidate review the materials and try again.

Introduction	Firefighters often perform emergency operations in the worst possible conditions. One condition that can be controlled to a certain degree is darkness. Firefighters can safely provide working light for fire fighting, rescue, and other emergency operations by using modern portable lighting equipment. It is important that firefighters know how to safely set up such equipment.
Equipment and Personnel	• Two firefighters in turnout gear • Portable power generator • Generator fuel • Portable floodlight • One portable light in telescoping stand • Three extension cords

- Junction box
- Appropriate equipment operation and service manuals
- "Operations area" as marked off or indicated by instructor

✔ **Note:** Although firefighters will commonly have access to a preconnected vehicle-mounted generator, this job sheet details the procedure for using a portable generator in order to allow candidates to practice proper lifting, placement, and connecting techniques.

Job Steps	Key Points	Attempt No. 1	2	3
1. Position portable light/ telescoping stand.	1. a. On dry, level surface	___	___	___
	b. In appropriate area to shed light on operations area	___	___	___
	c. In area where light will not "blind" working firefighters	___	___	___
	d. Out of main traffic area	___	___	___
	e. Handling without bulb breakage	___	___	___
	f. Unfolding legs fully and locking in place for sturdy support	___	___	___
	g. Adjusting to appropriate height and angle	___	___	___
2. Position portable floodlight.	2. a. On dry, level, elevated surface (a second firefighter may be positioned to hold this light)	___	___	___
	b. In appropriate area to shed light on operations arena	___	___	___
	c. In area where light will not "blind" working firefighters	___	___	___
	d. Out of main traffic area	___	___	___
	e. Handling without bulb breakage	___	___	___
3. Remove junction box from apparatus.	3. (None)	___	___	___
4. Position junction box.	4. a. Within reach of extension cords	___	___	___
	b. Out of main traffic area	___	___	___
5. Attach extension cord to each light.	5. (None)	___	___	___

Job Steps	Key Points	1	2	3
6. Attach other end of extension cords to junction box.	6. a. Comparing total wattage of load to capacity of junction box	—	—	—
	b. Without forcing plugs into junction box outlets	—	—	—
7. (Two firefighters) Remove portable generator from apparatus.	7. Lifting and carrying properly	—	—	—
8. Fill generator with fuel if necessary.	8. a. Checking fuel level	—	—	—
	b. Filling to proper level	—	—	—
	c. Avoiding spilling fuel or getting fuel on gloves or clothing	—	—	—
	d. Returning fuel container to appropriate apparatus storage	—	—	—
9. (Two firefighters) Position generator.	9. a. Within reach of junction box	—	—	—
	b. Out of main traffic area	—	—	—
10. Attach junction box to generator.	10. (None)	—	—	—
11. Turn on generator.	11. Using cord reel	—	—	—
12. Reposition lights as necessary.	12. a. On dry, level, elevated surface (a second firefighter may be positioned to hold a light)	—	—	—
	b. In appropriate area to shed light on operations arena	—	—	—
	c. In area where light will not "blind" working firefighters	—	—	—
	d. Out of main traffic area	—	—	—
	e. Handling without bulb breakage	—	—	—
13. Turn off generator.	13. Per manufacturer's instructions	—	—	—

Job Steps	Key Points	Attempt No. 1	2	3
14. Dismantle lighting equipment and return to proper storage.	14. a. Avoiding contact with hot bulbs	___	___	___
	b. Disconnecting cords at connections, not pulling on cords	___	___	___
	c. Rewinding cords	___	___	___
	d. *(Two firefighters)* Lifting generator properly	___	___	___
	Time (Total)	___	___	___

Evaluator's Comments _____

_____

_____

Job Sheet 7A-2
Service and Maintain Portable Power Plants and Lighting Equipment

Name _____ Date _____

Evaluator _____ Overall Competency Rating _____

References | NFPA 1001, Prevention, Preparedness, and Maintenance 4-5.2
Essentials, page 188

Prerequisites | None

Student's Instructions | To meet evaluation standards, you must perform this job within _____ *[amount of time, if applicable]*; you may have _____ attempts. When you are ready to perform this job, ask your instructor to observe the procedure and complete this form. To show mastery of this job, you must perform all steps to receive an overall competency rating of at least 2.

Competency Rating Scale

3 — Skilled — Meets all evaluation criteria and standards; performs task independently on first attempt; requires no additional practice or training.

2 — Moderately skilled — Meets all evaluation criteria and standards; performs task independently; additional practice is recommended.

1 — Unskilled — Is unable to perform the task; additional training required.

☒ **— Unassigned** — Job sheet task is not required or has not been performed.

✔ **Evaluator's Note:** Formulate and inform the candidate of the standards for this task (time allowed and number of attempts). Observe the candidate perform the task, check the step/key point under the appropriate attempt number as accomplished, record total time (if appropriate), and then use the rating scale above to assign an overall competency rating. If the candidate is unable to perform any step of this job, have the candidate review the materials and try again.

Introduction | Proper service and maintenance of portable power plants and lighting equipment will help to ensure their efficient and reliable operation. The failure of these systems can have disastrous consequences. For example, a search team may be left in the dark, a rescue squad may lose power to electric saws during extrication, or an attack team could misidentify structural members when ventilating a roof in poor light. For these reasons, firefighters must be certain that portable power plants and lighting equipment are operating at peak efficiency through proper maintenance.

Equipment and Personnel |
• One or two firefighters, depending on size of power plant
• Portable power plant (generator)
• Generator manufacturer's recommended fuel

- Generator manufacturer's recommended oil
- Lighting equipment
- Spark plug gap gauge
- Drain pan
- Gloves
- Shop cloth
- Lighting and power plant service and maintenance manuals
- Spare light bulbs appropriate to lights being tested

Job Steps	Key Points	Attempt No. 1	2	3
1. Remove generator from storage.	1. Lifting properly	—	—	—
2. Inspect generator spark plug.	2. a. Reporting damage, corrosion, or carbon accumulation	—	—	—
	b. Reporting cracks in porcelain	—	—	—
	c. Measuring gap for conformity to manufacturer's specifications	—	—	—
3. Inspect spark plug wire.	3. a. Tightening connection if too loose	—	—	—
	b. Reporting any frayed insulation found	—	—	—
4. Replace generator spark plug if inspection reveals damage or nonconformity.	4. With spark plug recommended by manufacturer and set to correct gap	—	—	—
5. Check generator carburetor.	5. Reporting any leaks found	—	—	—
6. Replace remaining fuel with fresh.	6. a. If fuel is three weeks old or older	—	—	—
	b. Discarding old fuel in approved manner and receptacle	—	—	—
7. Fill generator with fuel as necessary.	7. a. Checking fuel level	—	—	—
	b. Filling to manufacturer's recommended level	—	—	—
	c. Avoiding spilling fuel or getting fuel on gloves or clothing	—	—	—
	d. Returning fuel container to appropriate storage	—	—	—

Job Steps	Key Points	Attempt No. 1	2	3
8. Check generator oil level.	8. a. Withdrawing dipstick and wiping it clean	___	___	___
	b. Reinserting dipstick	___	___	___
	c. Withdrawing and reading dipstick markings	___	___	___
9. Replenish oil as necessary.	9. a. Using type and weight oil recommended by manufacturer	___	___	___
	b. Filling to level recommended by manufacturer	___	___	___
10. Inspect all electrical cords.	10. a. Reporting any frayed or damaged insulation	___	___	___
	b. Checking for missing or bent prongs	___	___	___
	c. Removing damaged cords from service	___	___	___
11. Test operation of lighting equipment.	11. a. Connecting one light at a time to generator	___	___	___
	b. Turning on generator	___	___	___
	c. Avoiding looking directly at lighted bulbs	___	___	___
12. Replace light bulbs as necessary.	12. a. Shutting off power before removing bulb	___	___	___
	b. Wearing gloves	___	___	___
	c. Avoiding burn injury	___	___	___
	d. Avoiding bulb breakage	___	___	___
13. Discard faulty bulbs.	13. In approved manner and receptacle	___	___	___
14. Clean work area, and return equipment to proper storage.	14. (None)	___	___	___

	Time (Total)	___	___	___

Evaluator's Comments _____

_____

_____

Job Sheet 7A-3
Use a Hydraulic Jack

Name _____ Date _____

Evaluator _____ Overall Competency Rating _____

References | NFPA 1001, Rescue Operations 4-4.2
Essentials, page 190

Prerequisites | None

Student's Instructions | To meet evaluation standards, you must perform this job within _____ *[amount of time, if applicable]*; you may have _____ attempts. When you are ready to perform this job, ask your instructor to observe the procedure and complete this form. To show mastery of this job, you must perform all steps to receive an overall competency rating of at least 2.

Competency Rating Scale

3 — Skilled — Meets all evaluation criteria and standards; performs task independently on first attempt; requires no additional practice or training.

2 — Moderately skilled — Meets all evaluation criteria and standards; performs task independently; additional practice is recommended.

1 — Unskilled — Is unable to perform the task; additional training required.

☒ **— Unassigned** — Job sheet task is not required or has not been performed.

✔ **Evaluator's Note:** Formulate and inform the candidate of the standards for this task (time allowed and number of attempts). Observe the candidate perform the task, check the step/key point under the appropriate attempt number as accomplished, record total time (if appropriate), and then use the rating scale above to assign an overall competency rating. If the candidate is unable to perform any step of this job, have the candidate review the materials and try again.

Introduction | The following job steps provide general procedures for operating a hydraulic jack. Be thoroughly familiar with the tool, its operating principles, methods, and limitations. *ALWAYS read and follow the manufacturer's directions and cautions before powering or operating a tool.*

!CAUTION: Before powering or operating the following tool, you must be dressed in full protective clothing and wearing eye protection.

Equipment and Personnel |
- Two or three firefighters (one to operate jack and one to place cribbing, with third necessary for operating a hand-operated hydraulic pump if used)
- Shims
- Cribbing and shoring blocks and wedges
- Vehicle-mounted, portable motor-driven, or hand-operated hydraulic pump
- Hydraulic jack
- Wrecked automobile(s) or other object(s) appropriate for skill demonstration

---

Job Steps	Key Points	Attempt No.		
		1	2	3
1. Assess the jack footing.	1. Solid, flat, level jacking surface	___	___	___
2. Adjust your protective clothing.	2. a. Faceshield lowered	___	___	___
	b. Gloves on	___	___	___
3. Place a flat board or steel plate under the load. !CAUTION: Do not reach under the load. A slipped load can cause crushing injury to the hands.	3. a. To level load	___	___	___
	b. Pushed under	___	___	___
	c. Where jack base will sit	___	___	___
	d. Shimmed level	___	___	___
4. Check the condition of jack's hydraulic hose and components. !CAUTION: Do not use tool if the hydraulic hose is damaged in any way.	4. a. Gasket in good condition or replaced as necessary	___	___	___
	b. Hose undamaged	___	___	___
5. Position the hydraulic pump.	5. Outside work area ✔ Note: If the hydraulic pump is hand-operated, it is located in the work area and operated by a second rescuer while the first rescuer operates the jack. The pump operator is responsible for checking the pump's hydraulic lines.	___	___	___
6. Connect jack's hydraulic hose to power source.	6. a. To outlet on hydraulic pump	___	___	___
	b. Snug fit	___	___	___
7. Position jack. !CAUTION: Do not reach under the load. A slipped load can cause crushing injury to the hands.	7. a. Under medium-duty load to be lifted	___	___	___
	b. Under solid area of load	___	___	___
	c. On level surface	___	___	___
	d. Retracted	___	___	___
	e. Held by sides or safely pushed under load	___	___	___
8. Turn the valve at the jack base.	8. Clockwise	___	___	___
9. Pump the jack lever.	9. a. To extend jack and lift the load	___	___	___
	b. Slowly	___	___	___

Job Steps	Key Points	1	2	3
10. Place cribbing. **!CAUTION:** Do not reach under the load. A slipped load can cause crushing injury to the hands.	10. Under the load	—	—	—
11. Build a crosshatch box formation.	11. a. As load is being elevated	—	—	—
	b. Holding cribbing and wedging pieces by their sides or handles	—	—	—
	c. Pushing in rear box portions with another piece of cribbing	—	—	—
	d. Wedging to provide maximum contact with load and box crib	—	—	—
	e. Using enough cribbing to support load	—	—	—
	f. Continually monitoring for any shifting	—	—	—
12. Jack up the load slightly.	12. a. To relieve load weight from cribbing	—	—	—
	b. When job is completed	—	—	—
13. Remove the cribbing blocks. **!CAUTION:** Do not reach under the load. A slipped load can cause crushing injury to the hands.	13. a. From top down	—	—	—
	b. By handles or sides	—	—	—
	c. Monitoring load continually for any shifting	—	—	—
14. Turn the valve at the jack base.	14. a. Counterclockwise	—	—	—
	b. To retract jack and lower the load	—	—	—
15. Pump the jack lever.	15. a. As cribbing is removed	—	—	—
	b. Slowly	—	—	—
16. Remove the jack.	16. a. Without reaching under load	—	—	—
	b. Monitoring load continually for shifting	—	—	—

Job Steps	Key Points	Attempt No.		
		1	2	3
17. Disconnect the jack from its power source.	17. Per manufacturer's instructions	—	—	—
18. Return the tool to proper storage.	18. Per manufacturer's instructions and department protocol	—	—	—
	Time (Total)	—	—	—

Evaluator's Comments _____

_____

_____

Job Sheet 7A-4
Use a Hydraulic Spreader

Name _____ **Date** _____

Evaluator _____ **Overall Competency Rating** _____

References	NFPA 1001, Rescue Operations 4-4.2 *Essentials,* page 189
Prerequisites	None
Student's Instructions	To meet evaluation standards, you must perform this job within _____ *[amount of time, if applicable]*; you may have _____ attempts. When you are ready to perform this job, ask your instructor to observe the procedure and complete this form. To show mastery of this job, you must perform all steps to receive an overall competency rating of at least 2.

Competency Rating Scale

3 — Skilled — Meets all evaluation criteria and standards; performs task independently on first attempt; requires no additional practice or training.

2 — Moderately skilled — Meets all evaluation criteria and standards; performs task independently; additional practice is recommended.

1 — Unskilled — Is unable to perform the task; additional training required.

☒ — **Unassigned** — Job sheet task is not required or has not been performed.

✔ **Evaluator's Note:** Formulate and inform the candidate of the standards for this task (time allowed and number of attempts). Observe the candidate perform the task, check the step/key point under the appropriate attempt number as accomplished, record total time (if appropriate), and then use the rating scale above to assign an overall competency rating. If the candidate is unable to perform any step of this job, have the candidate review the materials and try again.

Introduction	The following job steps provide general procedures for operating a hydraulic spreader. Be thoroughly familiar with the tool, its operating principles, methods, and limitations. *ALWAYS read and follow the manufacturer's directions and cautions before powering or operating a tool.* **!CAUTION:** Before powering or operating the following tool, you **must** be dressed in full protective clothing and wearing eye protection.
Equipment and Personnel	• Two firefighters (one to operate the spreader and one to operate a hand-operated hydraulic pump if used) • Safety goggles • Vehicle-mounted, portable motor-driven, or hand-operated hydraulic pump • Hydraulic spreader • Wrecked automobile(s) or other object(s) appropriate for skill demonstration

Job Steps	Key Points	Attempt No.		
		1	2	3
POWERING SPREADER WITH AUTOMATIC HYDRAULIC PUMP				
1. Check the condition of the hydraulic hoses and components.	1. a. Gaskets in good condition or replaced as necessary	——	——	——
	b. Hoses undamaged	——	——	——
2. Position the hydraulic pump.	2. a. Outside work area	——	——	——
	b. Within reach of spreader's hydraulic hoses	——	——	——
	c. Close enough to permit work to be accomplished without stretching hoses	——	——	——
3. Adjust your protective clothing.	3. a. Safety goggles on	——	——	——
	b. Faceshield lowered	——	——	——
	c. Gloves on	——	——	——
4. Connect the spreader's hydraulic hoses to the power source.	4. a. To outlets on hydraulic pump	——	——	——
	b. Snug connections	——	——	——
5. Pick up the spreader.	5. a. One hand on each arm handle *OR* one hand on arm handle and one on top bar grip	——	——	——
	b. Hydraulic lines over shoulder or to side and behind you	——	——	——
6. Close the spreader arms.	6. a. Button on top of the spreader moved in CLOSE direction	——	——	——
	b. With thumb (both hands on spreader handles)	——	——	——
	c. Until arms are fully closed	——	——	——
7. Insert the arms into the object to be spread or compressed. **!CAUTION:** Keep both hands on spreader handles and away from work to avoid crushing injury.	7. a. Tips only	——	——	——
	b. At proper angle for desired effect	——	——	——

Job Steps	Key Points	Attempt No. 1	2	3
8. Open the spreader arms.	8. a. Button on top of spreader moved in OPEN direction	___	___	___
	b. With thumb (both hands on spreader handles)	___	___	___
	c. Until desired compression or separation is achieved	___	___	___
9. Close the spreader arms.	9. Per Step 6	___	___	___
10. Remove the spreader from the work.	10. (None)	___	___	___
11. Disconnect the spreader from its power source.	11. Per manufacturer's instructions	___	___	___
12. Return the tool to proper storage.	12. a. Hydraulic hoses connected	___	___	___
	b. Per manufacturer's instructions and department protocol	___	___	___

Time (Total)   ___ ___ ___

Evaluator's Comments _____

_____

_____

POWERING SPREADER WITH MANUAL HYDRAULIC PUMP (OPERATED BY SECOND FIREFIGHTER) ✔ **Note:** It is recommended that a manually operated hydraulic pump be used only if an automatic pump is not available.				
(Spreader operator) 1. Check the condition of the spreader's hydraulic hoses and components.	1. a. Gaskets in good condition or replaced as necessary	___	___	___
	b. Hoses undamaged	___	___	___
(Pump operator) 2. Position the hydraulic pump.	2. a. Inside work area	___	___	___
	b. Out of way of spreader operator	___	___	___
	c. Within reach of hydraulic lines	___	___	___

Job Steps	Key Points	Attempt No.		
		1	2	3
	d. Close enough to permit work to be accomplished without stretching hoses	—	—	—
(Both firefighters) 3. Adjust your protective clothing.	3. a. Safety goggles on	—	—	—
	b. Faceshield lowered	—	—	—
	c. Gloves on	—	—	—
(Spreader operator) 4. Connect the spreader's hydraulic hoses to power source.	4. a. To outlet on hydraulic pump	—	—	—
	b. Snug connections	—	—	—
5. Pick up the spreader.	5. a. One hand on each handle arm *OR* one hand on handle arm and one on top bar grip	—	—	—
	b. Hydraulic lines over shoulder or to side and behind you	—	—	—
(Pump operator) 6. Turn the base lever on the hydraulic pump.	6. Counterclockwise	—	—	—
7. Close the spreader arms.	7. a. Pumping the hydraulic pump lever	—	—	—
	b. Until spreader operator signals that arms are fully closed	—	—	—
(Spreader operator) 8. Insert spreader arms into object to be spread or compressed. !CAUTION: Keep both hands on spreader handles and away from work to avoid crushing injury.	8. a. Tips only	—	—	—
	b. At proper angle for desired effect	—	—	—
(Pump operator) 9. Turn the base lever on the hydraulic pump.	9. Clockwise	—	—	—
10. Open the spreader arms.	10. a. Pumping the hydraulic pump lever	—	—	—
	b. Until spreader operator signals that desired compression or separation is achieved	—	—	—

Job Steps	Key Points	1	2	3
11. Repeat Steps 6 through 10.	11. Until desired cuts have been made	—	—	—
12. Close the spreader arms.	12. Per Steps 6 and 7	—	—	—
(Spreader operator) 13. Remove the spreader from the work.	13. (None)	—	—	—
14. Disconnect the spreader from its power source.	14. Per manufacturer's instructions	—	—	—
15. Return the tool to proper storage.	15. a. Hydraulic hoses connected	—	—	—
	b. Per manufacturer's instructions and department protocol	—	—	—
	Time (Total)	—	—	—

Evaluator's Comments _____

_____

_____

Job Sheet 7A-5
Use Hydraulic Shears

Name _____ Date _____

Evaluator _____ Overall Competency Rating _____

References	NFPA 1001, Rescue Operations 4-4.2 ***Essentials,*** page 189
Prerequisites	None
Student's Instructions	To meet evaluation standards, you must perform this job within _____ *[amount of time, if applicable]*; you may have _____ attempts. When you are ready to perform this job, ask your instructor to observe the procedure and complete this form. To show mastery of this job, you must perform all steps to receive an overall competency rating of at least 2.

> **Competency Rating Scale**
>
> **3 — Skilled** — Meets all evaluation criteria and standards; performs task independently on first attempt; requires no additional practice or training.
>
> **2 — Moderately skilled** — Meets all evaluation criteria and standards; performs task independently; additional practice is recommended.
>
> **1 — Unskilled** — Is unable to perform the task; additional training required.
>
> ☒ **—Unassigned** — Job sheet task is not required or has not been performed.
>
> ✔ **Evaluator's Note:** Formulate and inform the candidate of the standards for this task (time allowed and number of attempts). Observe the candidate perform the task, check the step/key point under the appropriate attempt number as accomplished, record total time (if appropriate), and then use the rating scale above to assign an overall competency rating. If the candidate is unable to perform any step of this job, have the candidate review the materials and try again.

Introduction	The following job steps provide general procedures for operating hydraulic shears. Be thoroughly familiar with the tool, its operating principles, methods, and limitations. *ALWAYS read and follow the manufacturer's directions and cautions before powering or operating a tool.* !CAUTION: Before powering or operating the following tool, you **must** be dressed in full protective clothing and wearing eye protection.
Equipment and Personnel	• Two firefighters (one to operate shears and one to operate a hand-operated hydraulic pump if used) • Safety goggles • Vehicle-mounted, portable motor-driven, or hand-operated hydraulic pump • Hydraulic shears • Wrecked automobile(s) or other object(s) appropriate for skill demonstration

Job Steps	Key Points	Attempt No.		
		1	**2**	**3**
POWERING SHEARS WITH AUTOMATIC HYDRAULIC PUMP				
1. Check the condition of the hydraulic hoses and components.	1. a. Gaskets in good condition or replaced as necessary	___	___	___
	b. Hoses undamaged	___	___	___
2. Position the hydraulic pump.	2. a. Outside work area	___	___	___
	b. Within reach of shears' hydraulic hoses	___	___	___
	c. Close enough to permit work to be accomplished without stretching hoses	___	___	___
3. Adjust your protective clothing.	3. a. Safety goggles on	___	___	___
	b. Faceshield lowered	___	___	___
	c. Gloves on	___	___	___
4. Connect the shears' hydraulic hoses to the power source.	4. a. To outlets on hydraulic pump	___	___	___
	b. Snug connection	___	___	___
5. Pick up the shears.	5. a. One hand on handle arm	___	___	___
	b. Other hand on top bar grip	___	___	___
	c. Hydraulic lines over shoulder or to side and behind you	___	___	___
6. Open the shears.	6. a. Operating button on top of shears in OPEN direction	___	___	___
	b. With thumb (both hands on handles)	___	___	___
	c. Until blades are fully open	___	___	___
7. Grasp the work to be cut with the shear blades. !CAUTION: Keep both hands on shear handles and away from work to avoid cutting/ crushing injury.	7. a. Work well back in blades	___	___	___
	b. At proper angle for desired cut	___	___	___
8. Close the shears.	8. a. Operating button on top of shears in CLOSE direction	___	___	___
	b. With thumb (both hands on shear handles)	___	___	___
	c. Until desired cut is made	___	___	___

Job Steps	Key Points	Attempt No. 1 2 3
9. Repeat Steps 6 through 8.	9. To make additional cuts as necessary	___ ___ ___
10. Remove the shears from the work.	10. (None)	___ ___ ___
11. Close the shear blades.	11. Per Step 8	___ ___ ___
12. Disconnect the shears from their power source.	12. Per manufacturer's instructions	___ ___ ___
13. Return the tool to proper storage.	13. a. Hydraulic hoses connected	___ ___ ___
	b. Per manufacturer's instructions and department protocol	___ ___ ___

Time (Total) ___ ___ ___

Evaluator's Comments _____

_____

_____

POWERING SHEARS WITH MANUALLY OPERATED HYDRAULIC PUMP

✔ **Note:** It is recommended that a manually operated hydraulic pump be used only if an automatic pump is not available.

Job Steps	Key Points	Attempt No.
(Shears operator)		
1. Check the condition of the shears' hydraulic hoses and components.	1. a. Gaskets in good condition or replaced as necessary	___ ___ ___
	b. Hoses undamaged in any way	___ ___ ___
(Pump operator)		
2. Position the hydraulic pump.	2. a. Inside work area	___ ___ ___
	b. Out of way of shears operator	___ ___ ___
	c. Within reach of hydraulic lines	___ ___ ___
	d. Close enough to permit work to be accomplished without stretching hoses	___ ___ ___

Job Steps	Key Points	Attempt No. 1 2 3

(Both firefighters)
3. Adjust your protective clothing.

3. a. Safety goggles on _ _ _
   b. Faceshield lowered _ _ _
   c. Gloves on _ _ _

(Shears operator)
4. Connect the shears' hydraulic hoses to power source.

4. a. To outlet on hydraulic pump _ _ _
   b. Snug connections _ _ _

5. Pick up the shears.

5. a. One hand on handle arm _ _ _
   b. Other hand on top bar grip _ _ _
   c. Hydraulic lines over shoulder or to side and behind you _ _ _

(Pump operator)
6. Turn the base lever on the hydraulic pump.

6. Clockwise _ _ _

7. Open the shears.

7. a. Pumping hydraulic pump lever _ _ _
   b. Until shears operator signals that arms are fully opened _ _ _

(Shears operator)
8. Grasp the work to be cut with the shear blades.

   !CAUTION: Keep both hands on shears' handles and away from work to avoid cutting/crushing injury.

8. a. Work well back in blades _ _ _
   b. At proper angle for desired cut _ _ _

(Pump operator)
9. Turn the base lever on the hydraulic pump.

9. Counterclockwise _ _ _

10. Close the shears.

10. a. Pumping hydraulic pump lever _ _ _
    b. Until shears operator signals that desired cut is made _ _ _

11. Repeat Steps 6 through 10.

11. Until desired cuts have been made _ _ _

(Shears operator)
12. Remove the shears from the work.

12. (None) _ _ _

13. Disconnect the shears from their power source.

13. Per manufacturer's instructions _ _ _

Job Steps	Key Points	Attempt No.		
		1	2	3
14. Return the tool to proper storage.	14. a. Hydraulic hoses connected	—	—	—
	b. Per manufacturer's instructions and department protocal	—	—	—
	Time (Total)	—	—	—

Evaluator's Comments _____

_____

_____

Job Sheet 7A-6
Use a Hydraulic Extension Ram

Name _____ Date _____

Evaluator _____ Overall Competency Rating _____

References | NFPA 1001, Rescue Operations 4-4.2
Essentials, page 189

Prerequisites | None

Student's Instructions | To meet evaluation standards, you must perform this job within _____ *[amount of time, if applicable]*; you may have _____ attempts. When you are ready to perform this job, ask your instructor to observe the procedure and complete this form. To show mastery of this job, you must perform all steps to receive an overall competency rating of at least 2.

Competency Rating Scale

3 — Skilled — Meets all evaluation criteria and standards; performs task independently on first attempt; requires no additional practice or training.

2 — Moderately skilled — Meets all evaluation criteria and standards; performs task independently; additional practice is recommended.

1 — Unskilled — Is unable to perform the task; additional training required.

☒ **— Unassigned** — Job sheet task is not required or has not been performed.

✔ **Evaluator's Note:** Formulate and inform the candidate of the standards for this task (time allowed and number of attempts). Observe the candidate perform the task, check the step/key point under the appropriate attempt number as accomplished, record total time (if appropriate), and then use the rating scale above to assign an overall competency rating. If the candidate is unable to perform any step of this job, have the candidate review the materials and try again.

Introduction | The following job steps provide general procedures for operating a hydraulic extension ram. Be thoroughly familiar with the tool, its operating principles, methods, and limitations. *ALWAYS read and follow the manufacturer's directions and cautions before powering or operating a tool.*

!CAUTION: Before powering or operating the following tool, you must be dressed in full protective clothing and wearing eye protection.

Equipment and Personnel |
- Two firefighters (one to operate ram and one to operate a hand-operated hydraulic pump if used)
- Vehicle-mounted, portable motor-driven, or hand-operated hydraulic pump
- Hydraulic ram
- Wrecked automobile(s) or other object(s) appropriate for skill demonstration

POWERING RAM WITH AUTOMATIC HYDRAULIC PUMP

✔ **Note:** It is recommended that a manually operated hydraulic pump be used only if an automatic pump is not available.

1. Check the condition of the hydraulic hoses and components.

    1. a. Gaskets in good condition or replaced as necessary   ___ ___ ___
        b. Hoses undamaged   ___ ___ ___

2. Position the hydraulic pump.

    2. a. Outside work area   ___ ___ ___
        b. Within reach of ram's hydraulic hoses   ___ ___ ___
        c. Close enough to permit work to be accomplished without stretching hoses   ___ ___ ___

3. Adjust your protective clothing.

    3. a. Safety goggles on   ___ ___ ___
        b. Faceshield lowered   ___ ___ ___
        c. Gloves on   ___ ___ ___

4. Connect the ram's hydraulic hoses to the power source.

    4. a. To outlets on hydraulic pump   ___ ___ ___
        b. Snug connection   ___ ___ ___

5. Pick up and position hydraulic ram.
    !CAUTION: Keep both hands on ram's handles and away from ram's telescoping arms to avoid crushing injury.

    5. a. By handle(s) only   ___ ___ ___
        b. Securely between objects to be separated or compressed   ___ ___ ___
        c. Hydraulic lines not tangled   ___ ___ ___
        d. Hydraulic lines on outside of work   ___ ___ ___

6. Extend ram's telescoping arms.

    6. a. When ram is effectively positioned for work   ___ ___ ___
        b. Operating button on top of ram in EXTEND direction   ___ ___ ___
        c. With thumb (both hands clear of arms)   ___ ___ ___
        d. Until desired separation or compression is achieved   ___ ___ ___

7. Stop extension of the ram's telescoping arms.

    7. a. When desired extension or compression is achieved   ___ ___ ___

Job Steps	Key Points	Attempt No.		
		1	2	3
	b. Operating button on top of ram in OFF or NEUTRAL position	___	___	___
	c. With thumb (both hands clear of arms)	___	___	___
	d. Until telescoping arms stop extending	___	___	___
8. Retract ram's telescoping arms.	8. a. When work is completed	___	___	___
	b. Operating button on top of ram in RETRACT direction	___	___	___
	b. With thumb (both hands clear of arms)	___	___	___
	c. Until telescoping arms are fully retracted	___	___	___
9. Disconnect ram from power source.	9. Per manufacturer's instructions	___	___	___
10. Remove ram from work.	10. By handle(s)	___	___	___
11. Return tool to proper storage.	11. a. Hydraulic hoses connected	___	___	___
	b. Per manufacturer's instructions and department protocol	___	___	___
	Time (Total)	___	___	___

Evaluator's Comments _____

_____

_____

POWERING RAM WITH MANUALLY OPERATED HYDRAULIC PUMP ✔ **Note:** It is recommended that a manually operated hydraulic pump be used only if an automatic pump is not available. *(Ram operator)* 1. Check the condition of the ram's hydraulic hoses and components.	1. a. Gaskets in good condition or replaced as necessary ___ ___ ___ b. Hoses undamaged ___ ___ ___

Job Steps	Key Points	1	2	3
(Pump operator)				
2. Position the hydraulic pump.	2. a. Inside work area	___	___	___
	b. Out of way of ram operator	___	___	___
	c. Within reach of hydraulic lines	___	___	___
	d. Close enough to permit work to be accomplished without stretching hoses	___	___	___
(Both firefighters)				
3. Adjust your protective clothing.	3. a. Safety goggles on	___	___	___
	b. Faceshield lowered	___	___	___
	c. Gloves on	___	___	___
(Ram operator)				
4. Connect the ram's hydraulic hoses to power source.	4. a. To outlet on hydraulic pump	___	___	___
	b. Snug connections	___	___	___
5. Pick up and position hydraulic ram. **!CAUTION:** Keep both hands on ram's handles and away from ram's telescoping arms to avoid crushing injury.	5. a. By handle(s) only	___	___	___
	b. Securely between objects to be separated or compressed	___	___	___
	c. Hydraulic lines not tangled	___	___	___
	d. Hydraulic lines on outside of work	___	___	___
(Pump operator)				
6. Extend the ram's telescoping arms.	6. a. When ram is effectively positioned for work	___	___	___
	b. Turning base lever on hydraulic pump clockwise	___	___	___
	c. Pumping hydraulic pump lever	___	___	___
	d. Until ram operator signals that desired separation or compression is achieved	___	___	___
7. Retract the ram's telescoping arms.	7. a. When job is completed	___	___	___
	b. Turning base lever on hydraulic pump counter-clockwise	___	___	___
	c. Pumping hydraulic pump lever	___	___	___
	d. Until telescoping arms are fully retracted	___	___	___

Job Steps	Key Points	Attempt No. 1 2 3

(Ram operator)

8. Remove the ram from the work.

9. Disconnect the ram from its power source.

10. Return the tool to proper storage.

8. By its handle(s)   —— —— ——

9. Per manufacturer's instructions   —— —— ——

10. a. Hydraulic hoses connected   —— —— ——

    b. Per manufacturer's instructions and department protocol   —— —— ——

| Time (Total) | —— —— —— |

Evaluator's Comments _____

_____

_____

Job Sheet 7A-7
Use a Bar Screw Jack

Name _____ Date _____

Evaluator _____ Overall Competency Rating _____

References	NFPA 1001, Rescue Operations 4-4.2 *Essentials,* page 191
Prerequisites	None
Student's Instructions	To meet evaluation standards, you must perform this job within _____ *[amount of time, if applicable]*; you may have _____ attempts. When you are ready to perform this job, ask your instructor to observe the procedure and complete this form. To show mastery of this job, you must perform all steps to receive an overall competency rating of at least 2.

> **Competency Rating Scale**
>
> **3 — Skilled** — Meets all evaluation criteria and standards; performs task independently on first attempt; requires no additional practice or training.
>
> **2 — Moderately skilled** — Meets all evaluation criteria and standards; performs task independently; additional practice is recommended.
>
> **1 — Unskilled** — Is unable to perform the task; additional training required.
>
> ☒ — **Unassigned** — Job sheet task is not required or has not been performed.
>
> ✔ **Evaluator's Note:** Formulate and inform the candidate of the standards for this task (time allowed and number of attempts). Observe the candidate perform the task, check the step/key point under the appropriate attempt number as accomplished, record total time (if appropriate), and then use the rating scale above to assign an overall competency rating. If the candidate is unable to perform any step of this job, have the candidate review the materials and try again.

Introduction	The following job steps provide general procedures for operating a bar screw jack. Be thoroughly familiar with the tool, its operating principles, methods, and limitations. *ALWAYS read and follow the manufacturer's directions and cautions before powering or operating a tool.* **!CAUTION:** Before operating the following tool, you must be dressed in full protective clothing and wearing eye protection.
Equipment and Personnel	• One firefighter in full protective clothing • Bar screw jack • Flat board(s) or steel plate (as necessary for jack base) • Wrecked automobile or other heavy objects or situation requiring shoring or compression

Job Steps	Key Points	Attempt No.		
		1	2	3
1. Assess jack footing.	1. Solid, flat jacking surface	—	—	—
2. Adjust protective clothing.	2. a. Faceshield lowered	—	—	—
	b. Gloves on	—	—	—
3. Adjust jack footing as necessary.	3. Flat board or steel plate where jack will exert pressure	—	—	—
4. Position jack.	4. a. Between solid members to be shored or held under compression	—	—	—
	b. Not under a load to be lifted	—	—	—
	c. Jack core retracted to allow bar insertion	—	—	—
5. Insert jack bar into jack.	5. Into hole at top of jack core	—	—	—
6. Extend jack core.	6. a. To apply pressure	—	—	—
	b. Turning bar clockwise	—	—	—
	c. Until sufficient tension is produced	—	—	—
7. Remove jack bar.	7. Pulled from hole in top of jack core	—	—	—
8. Prepare to remove the jack.	8. a. Reinserting jack bar into hole in top of jack core	—	—	—
	b. When compression or shoring is no longer needed	—	—	—
9. Retract the jack.	9. a. Turning bar counterclockwise	—	—	—
	b. Slowly	—	—	—
	c. Until core retracts enough to remove jack	—	—	—
10. Return jack to proper storage.	10. Per manufacturer's instructions and department protocol	—	—	—
	Time (Total)	—	—	—

Evaluator's Comments _____

_____

_____

Job Sheet 7A-8
Use a Trench Screw Jack

Name _____ Date _____

Evaluator _____ Overall Competency Rating _____

References	NFPA 1001, Rescue Operations 4-4.2 *Essentials,* page 191
Prerequisites	None
Student's Instructions	To meet evaluation standards, you must perform this job within _____ *[amount of time, if applicable]*; you may have _____ attempts. When you are ready to perform this job, ask your instructor to observe the procedure and complete this form. To show mastery of this job, you must perform all steps to receive an overall competency rating of at least 2.

> **Competency Rating Scale**
>
> **3 — Skilled** — Meets all evaluation criteria and standards; performs task independently on first attempt; requires no additional practice or training.
>
> **2 — Moderately skilled** — Meets all evaluation criteria and standards; performs task independently; additional practice is recommended.
>
> **1 — Unskilled** — Is unable to perform the task; additional training required.
>
> ⊠ **— Unassigned** — Job sheet task is not required or has not been performed.
>
> ✔ **Evaluator's Note:** Formulate and inform the candidate of the standards for this task (time allowed and number of attempts). Observe the candidate perform the task, check the step/key point under the appropriate attempt number as accomplished, record total time (if appropriate), and then use the rating scale above to assign an overall competency rating. If the candidate is unable to perform any step of this job, have the candidate review the materials and try again.

| **Introduction** | The following job steps provide general procedures for operating a trench screw jack and placing shoring. Be thoroughly familiar with the tool, its operating principles, methods, and limitations. *ALWAYS read and follow the manufacturer's directions and cautions before powering or operating a tool.*

✔ **Note:** This job sheet provides practice in operating trench screw jacks. These tools are commonly used in trench rescue operations. However, the purpose of this job sheet is to provide training in the use of trench screw jacks, not in performing trench rescue operations. Additional training will be required to qualify the candidate for trench rescues.

!CAUTION: Before operating the following tool, you must be dressed in full protective clothing and wearing eye protection. |

Equipment and Personnel
- Two firefighters (one to operate jack and one to place shoring) in full protective clothing and helmets with faceshields
- Two trench screw jacks
- Boards to use as shoring
- Steel pipe of appropriate diameter for jack and appropriate length for demonstration area
- Hallway or other area to simulate a trench
- Padding (optional)

✔ **Note:** Your instructor may require you to place padding between the shoring and the demonstration area prior to extending jacks.

Job Steps	Key Points	Attempt No. 1	2	3
1. Position shoring boards.	1. a. Placing shoring against trench sidewalls parallel to each other	___	___	___
	b. Placing shoring vertically	___	___	___
2. Assemble one trench screw jack.	2. a. Choosing appropriate diameter steel pipe approximately 8 inches *(200 mm)* shorter than the distance between the shoring boards	___	___	___
	b. Installing swivel footplate in one end of pipe and a swivel footplate with a threaded stem in the other end of the pipe	___	___	___
3. Position trench jack for top brace.	3. a. Positioning jack no more than 1.5 feet *(0.5 m)* from the top of the trench	___	___	___
	b. Placing jack footplate without threaded stem against one shoring board	___	___	___
4. Tighten jack.	4. a. Rotating adjusting nut counterclockwise to lengthen jack	___	___	___
	b. Placing footplates squarely against shoring braces	___	___	___
	c. Tightening until there is enough pressure for jack to be self-supporting without deforming braces	___	___	___

Job Steps	Key Points	Attempt No.		
		1	2	3
5. Position trench jack for bottom brace.	5. a. Positioning jack no more than 2 feet *(0.6 m)* from the floor of the trench	—	—	—
	b. Placing jack footplate without threaded stem against one shoring board	—	—	—
6. Tighten jack.	6. a. Rotating adjusting nut counterclockwise to lengthen jack	—	—	—
	b. Placing footplates squarely against shoring braces	—	—	—
	c. Tightening until there is enough pressure for jack to be self-supporting without deforming braces	—	—	—
7. Adjust jacks.	7. a. Tightening or loosening upper jack for proper pressure	—	—	—
	b. Tightening or loosening lower jack for proper pressure	—	—	—
8. Remove lower jack.	8. a. Rotating adjusting nut clockwise to shorten jack	—	—	—
	b. Removing and disassembling jack	—	—	—
9. Remove upper jack.	9. a. Rotating adjusting nut clockwise to shorten jack	—	—	—
	b. Removing and disassembling jack	—	—	—
10. Remove shoring boards.	10. (None)	—	—	—
11. Return jacks and other components to proper storage.	11. Per manufacturer's instructions and department protocol	—	—	—
	Time (Total)	—	—	—

Evaluator's Comments _____

_____

_____

Job Sheet 7A-9
Use a Ratchet-Lever Jack

Name _____ Date _____

Evaluator _____·Overall Competency Rating _____

References | NFPA 1001, Rescue Operations 4-4.2
Essentials, page 191

Prerequisites | None

Student's Instructions | To meet evaluation standards, you must perform this job within _____ *[amount of time, if applicable]*; you may have _____ attempts. When you are ready to perform this job, ask your instructor to observe the procedure and complete this form. To show mastery of this job, you must perform all steps to receive an overall competency rating of at least 2.

Competency Rating Scale

3 — Skilled — Meets all evaluation criteria and standards; performs task independently on first attempt; requires no additional practice or training.

2 — Moderately skilled — Meets all evaluation criteria and standards; performs task independently; additional practice is recommended.

1— Unskilled — Is unable to perform the task; additional training required.

☒ — Unassigned — Job sheet task is not required or has not been performed.

✔ **Evaluator's Note:** Formulate and inform the candidate of the standards for this task (time allowed and number of attempts). Observe the candidate perform the task, check the step/key point under the appropriate attempt number as accomplished, record total time (if appropriate), and then use the rating scale above to assign an overall competency rating. If the candidate is unable to perform any step of this job, have the candidate review the materials and try again.

Introduction | The following job steps provide general procedures for operating a ratchet-lever jack and placing cribbing. Be thoroughly familiar with the tool, its operating principles, methods, and limitations. *ALWAYS read and follow the manufacturer's directions and cautions before powering or operating a tool.*

!CAUTION: Before operating the following tool, you must be dressed in full protective clothing and wearing eye protection.

Equipment and Personnel |
- Two firefighters (one to operate jack and one to place cribbing)
- Ratchet-lever jack
- Flat board or steel plate (as necessary for jack base)
- Shims
- Cribbing and shoring blocks and wedges
- Wrecked automobile(s) or other heavy object(s) appropriate for skill demonstration

Job Steps	Key Points	1	2	3
1. Assess jack footing.	1. Solid, flat, level jacking surface	—	—	—
2. Adjust protective clothing.	2. a. Faceshield lowered	—	—	—
	b. Gloves on	—	—	—
3. Level jack footing as necessary. !CAUTION: Do not reach under the load. A slipped load can cause crushing injury to the hands.	3. a. Flat board or steel plate *pushed* under load	—	—	—
	b. Where jack base will sit	—	—	—
	c. Shimmed level	—	—	—
4. Position jack. !CAUTION: Do not reach under the load. A slipped load can cause crushing injury to the hands.	4. a. Under medium-duty load to be lifted	—	—	—
	b. Under solid area of load	—	—	—
	c. On level board or plate as necessary			
	d. Carriage retracted	—	—	—
	e. Held by sides or safely pushed under load	—	—	—
5. Set carriage lever.	5. In UP position	—	—	—
6. Lift the load.	6. a. Pumping jack lever	—	—	—
	b. Slowly	—	—	—
7. Place cribbing under the load. !CAUTION: Do not reach under the load. A slipped load can cause crushing injury to the hands.	7. a. Not reaching under load	—	—	—
	b. Building a crosshatch box formation	—	—	—
	c. As load is being elevated	—	—	—
	d. Holding cribbing and wedging pieces by their sides or handles	—	—	—
	e. Pushing in rear box portions with another piece of cribbing	—	—	—
	f. Wedging to provide maximum contact with load and box crib	—	—	—
	g. Using enough cribbing to support load	—	—	—
	h. Continually monitoring for any shifting	—	—	—

Job Steps	Key Points	Attempt No. 1	2	3
8. Move weight of load slightly off cribbing.	8. a. When job is complete	——	——	——
	b. Carriage lever in UP position	——	——	——
	c. Pumping jack lever	——	——	——
	d. Slowly	——	——	——
9. Remove cribbing blocks. **!CAUTION:** Do not reach under the load. A slipped load can cause crushing injury to the hands.	9. a. From top down	——	——	——
	b. By handles or sides	——	——	——
	c. Monitoring continually for any shifting	——	——	——
10. Set carriage lever.	10. In DOWN position	——	——	——
11. Pump jack lever.	11. Slowly	——	——	——
12. Remove jack.	12. a. Slowly	——	——	——
	b. As cribbing is removed	——	——	——
13. Return jack to proper storage.	13. Per manufacturer's instructions and department protocol	——	——	——
	Time (Total)	——	——	——

Evaluator's Comments _____

_____

_____

Job Sheet 7A-10
Use a Pneumatic Chisel/Hammer

Name _____ Date _____

Evaluator _____ Overall Competency Rating _____

References	NFPA 1001, Rescue Operations 4-4.2 *Essentials,* page 193
Prerequisites	None
Student's Instructions	To meet evaluation standards, you must perform this job within _____ *[amount of time, if applicable]*; you may have _____ attempts. When you are ready to perform this job, ask your instructor to observe the procedure and complete this form. To show mastery of this job, you must perform all steps to receive an overall competency rating of at least 2.

Competency Rating Scale

3 — Skilled — Meets all evaluation criteria and standards; performs task independently on first attempt; requires no additional practice or training.

2 — Moderately skilled — Meets all evaluation criteria and standards; performs task independently; additional practice is recommended.

1 — Unskilled — Is unable to perform the task; additional training required.

☒ **— Unassigned** — Job sheet task is not required or has not been performed.

✔ **Evaluator's Note:** Formulate and inform the candidate of the standards for this task (time allowed and number of attempts). Observe the candidate perform the task, check the step/key point under the appropriate attempt number as accomplished, record total time (if appropriate), and then use the rating scale above to assign an overall competency rating. If the candidate is unable to perform any step of this job, have the candidate review the materials and try again.

Introduction	The following job steps provide general procedures for operating a pneumatic chisel/hammer. Be thoroughly familiar with the tool, its operating principles, methods, and limitations. *ALWAYS read and follow the manufacturer's directions and cautions before powering or operating a tool.* **!CAUTION:** Before powering or operating the following tool, you ***must*** be dressed in full protective clothing and wearing eye protection.
Equipment and Personnel	• One firefighter in full protective clothing • Safety goggles • Portable air compressor or SCBA or cascade cylinders • Pneumatic chisel • Wrecked automobile(s) or other object(s) appropriate for skill demonstration

---

Job Steps	Key Points	Attempt No. 1 2 3

Job Steps	Key Points			

1. Adjust your protective clothing.

   1. a. Safety goggles on   ___ ___ ___

        b. Faceshield lowered   ___ ___ ___

        c. Gloves on   ___ ___ ___

2. Use or attach appropriate cutting bit.

   2. a. Inspected for defects   ___ ___ ___

        b. Replaced if damaged or flawed   ___ ___ ___

        c. Chucked firmly in tool   ___ ___ ___

3. Check area in which work is to be done.

   !CAUTION: Contact between the cutting bit and metal may produce sparks that will ignite flammable vapors. Never use on containers that have contained flammable materials.

   3. a. No ignition hazards   ___ ___ ___

        b. No flammable vapors   ___ ___ ___

4. Check condition of pneumatic hose and components.

   4. a. Replacing hose gaskets as necessary   ___ ___ ___

        b. Not using tool if hose is damaged in any way   ___ ___ ___

5. Connect pneumatic hose to chisel.

   5. a. Working end to base of chisel handle   ___ ___ ___

        b. Snug connection   ___ ___ ___

        c. Following manufacturer's directions   ___ ___ ___

6. Connect chisel to power source.

   6. a. Pressure gauges connected to high-pressure outlet of air supply (if SCBA or cascade bottle)   ___ ___ ___

        b. Regulator end of pneumatic hose connected to compressor outlet *or* to pressure gauge outlet (if SCBA or cascade bottle)   ___ ___ ___

7. Adjust pressure gauge valve.

   7. a. If not preset   ___ ___ ___

        b. To operating pressure specified by chisel manufacturer   ___ ___ ___

Job Steps	Key Points	Attempt No. 1	2	3
8. Pick up chisel.	8. a. One hand on pistol grip	——	——	——
	b. Other hand supporting chisel barrel	——	——	——
	c. Pneumatic line untangled	——	——	——
	d. Pneumatic line over shoulder or to side and behind you	——	——	——
9. Position chisel effectively to work.	9. a. Bit in contact with work	——	——	——
	b. Bit at appropriate angle to accomplish desired cut	——	——	——
10. Operate chisel to start cut. !CAUTION: Keep both hands on chisel handle and body and away from chisel's rotating bit to avoid injury.	10. a. Trigger held in depressed position with index finger	——	——	——
	b. Both hands on chisel	——	——	——
	c. Bit not forced through work	——	——	——
11. Reposition chisel bit as necessary.	11. a. Finger released from trigger	——	——	——
	b. Bit placed in contact with new site at desired angle	——	——	——
12. Stop chisel bit.	12. a. Finger released from trigger	——	——	——
	b. When cut has been made	——	——	——
13. Disconnect chisel from power source.	13. a. Compressor or pressure gauge valve on air bottle shut off	——	——	——
	b. Pneumatic line bled (trigger depressed)	——	——	——
	c. Pneumatic line removed from chisel	——	——	——
	d. Pneumatic line removed from air source	——	——	——
14. Return tool to proper storage.	14. Per manufacturer's instructions and department protocol	——	——	——
	Time (Total)	——	——	——

Evaluator's Comments _____

_____

_____

Job Sheet 7A-11
Use a Truck-Mounted Winch

Name _____ **Date** _____

Evaluator _____ **Overall Competency Rating** _____

References	NFPA 1001, Rescue Operations 4-4.2 *Essentials,* page 194
Prerequisite	Job Sheet 7A-14 — Use a Block and Tackle
Student's Instructions	To meet evaluation standards, you must perform this job within _____ *[amount of time, if applicable]*; you may have _____ attempts. When you are ready to perform this job, ask your instructor to observe the procedure and complete this form. To show mastery of this job, you must perform all steps to receive an overall competency rating of at least 2.

Competency Rating Scale

3 — Skilled — Meets all evaluation criteria and standards; performs task independently on first attempt; requires no additional practice or training.

2 — Moderately skilled — Meets all evaluation criteria and standards; performs task independently; additional practice is recommended.

1 — Unskilled — Is unable to perform the task; additional training required.

☒ —**Unassigned** — Job sheet task is not required or has not been performed.

✔ **Evaluator's Note:** Formulate and inform the candidate of the standards for this task (time allowed and number of attempts). Observe the candidate perform the task, check the step/key point under the appropriate attempt number as accomplished, record total time (if appropriate), and then use the rating scale above to assign an overall competency rating. If the candidate is unable to perform any step of this job, have the candidate review the materials and try again.

Introduction	The following job steps provide general procedures for operating a truck-mounted winch. Be thoroughly familiar with the tool, its operating principles, methods, and limitations. *ALWAYS read and follow the manufacturer's directions and cautions before powering or operating a tool.* !CAUTION: Before powering or operating the following tool, you must be dressed in full protective clothing and wearing eye protection.
Equipment and Personnel	• Rescue team of at least two firefighters in full protective clothing • Heavy-duty twine • Truck-mounted winch • Block and tackle with appropriate length, type, and diameter rope • Wrecked automobile(s) or other heavy object(s) appropriate for skill demonstration

---

Job Steps	Key Points	Attempt No.		
		1	2	3
1. Inspect the winch and cable.	1. Reporting any flaws or damage	___	___	___
2. Position the winch truck.	2. a. In line with object	___	___	___
	b. As close to object as practical	___	___	___
3. Adjust your protective clothing.	3. a. Faceshield lowered	___	___	___
	b. Gloves on	___	___	___
4. Attach the winch cable to the load.	4. a. Per manufacturer's directions and cautions	___	___	___
	b. Without removing windings from bottom layer of winch spool	___	___	___
5. Attach the block and tackle to the load for added safety.	5. a. Per block and tackle steps in Job Sheet 7A-14	___	___	___
	b. As close as possible to winch attachment point	___	___	___
6. Direct the pulling process. !CAUTION: A broken chain or cable can whiplash with lethal force.	6. a. From area out of danger of whiplash	___	___	___
	b. *(Truck)* With handheld, remote control operating device	___	___	___
	c. *(Winch fall line team)* Verbally	___	___	___
7. Release the tension on the winch cable.	7. a. When job is completed	___	___	___
	b. Per manufacturer's directions and cautions	___	___	___
8. Remove block and tackle from the load.	8. a. After all tension has been released from winch cable	___	___	___
	b. Per steps in Job Sheet 7A-14	___	___	___
9. Remove the winch hook from the load.	9. (None)	___	___	___
10. Rewind the winch cable.	10. Per manufacturer's directions and cautions	___	___	___

Time (Total)	___	___	___

Evaluator's Comments _____

_____

_____

Job Sheet 7A-12
Use a Come-Along

Name _____ Date _____

Evaluator _____ Overall Competency Rating _____

References	NFPA 1001, Rescue Operations 4-4.2 *Essentials,* page 194
Prerequisites	None
Student's Instructions	To meet evaluation standards, you must perform this job within _____ *[amount of time, if applicable]*; you may have _____ attempts. When you are ready to perform this job, ask your instructor to observe the procedure and complete this form. To show mastery of this job, you must perform all steps to receive an overall competency rating of at least 2.

> **Competency Rating Scale**
>
> **3 — Skilled** — Meets all evaluation criteria and standards; performs task independently on first attempt; requires no additional practice or training.
>
> **2 — Moderately skilled** — Meets all evaluation criteria and standards; performs task independently; additional practice is recommended.
>
> **1 — Unskilled** — Is unable to perform the task; additional training required.
>
> ☒ — **Unassigned** — Job sheet task is not required or has not been performed.
>
> ✔ **Evaluator's Note:** Formulate and inform the candidate of the standards for this task (time allowed and number of attempts). Observe the candidate perform the task, check the step/key point under the appropriate attempt number as accomplished, record total time (if appropriate), and then use the rating scale above to assign an overall competency rating. If the candidate is unable to perform any step of this job, have the candidate review the materials and try again.

Introduction	The following job steps provide general procedures for operating a come-along. Be thoroughly familiar with the tool, its operating principles, methods, and limitations. *ALWAYS read and follow the manufacturer's directions and cautions before powering or operating a tool.* **!CAUTION:** Before operating the following tool, you must be dressed in full protective clothing and wearing eye protection.
Equipment and Personnel	• One firefighter in full protective clothing • Come-along • Heavy-duty twine • Wrecked automobile(s) or other heavy object(s) appropriate for skill demonstration

Job Steps	Key Points	Attempt No. 1 2 3

Job Steps

1. Adjust protective clothing.

2. Anchor come-along.

3. Mouse or lock the cable hook.

4. Attach come-along.

5. Set drum latch.

6. Operate drum lever.

   !CAUTION: A broken chain or cable can whiplash with lethal force.

7. Reset drum latch.

Key Points

1. a. Faceshield lowered
   b. Gloves on

2. a. Drum/lever end
   b. To stationary object that will hold pull load
   c. In line of desired movement

3. *(Locking safety)* Closing safety latch
   (Mousing)
   a. By wrapping strong twine around hook
   b. Eight to ten times
   c. Several additional turns around mousing
   d. Ends tied securely

4. a. To object to be moved
   b. To part of object that can withstand pull load
   c. Using sufficient length of cable
   d. Hook moused or locked

5. In PULL position

6. a. To pull the object
   b. From behind come-along drum
   c. Back and forth
   d. Slowly and cautiously
   e. Until object is moved

7. a. When job is complete
   b. In RELEASE position

Job Steps	Key Points	Attempt No.		
		1	2	3
8. Operate drum lever.	8. a. Back and forth	___	___	___
	b. Until tension is released	___	___	___
9. Unhook come-along.	9. a. From moved object	___	___	___
	b. From anchor object	___	___	___
10. Return come-along to proper storage.	10. Per manufacturer's instructions and department protocol	___	___	___
	Time (Total)	___	___	___

Evaluator's Comments _____

_____

_____

Job Sheet 7A-13
Use Air Lifting Bag(s)

Name _____ Date _____

Evaluator _____ Overall Competency Rating _____

References | NFPA 1001, Rescue Operations 4-4.2
Essentials, pages 195, 196

Prerequisites | None

Student's Instructions | To meet evaluation standards, you must perform this job within _____ *[amount of time, if applicable]*; you may have _____ attempts. When you are ready to perform this job, ask your instructor to observe the procedure and complete this form. To show mastery of this job, you must perform all steps to receive an overall competency rating of at least 2.

> **Competency Rating Scale**
>
> **3 — Skilled** — Meets all evaluation criteria and standards; performs task independently on first attempt; requires no additional practice or training.
>
> **2 — Moderately skilled** — Meets all evaluation criteria and standards; performs task independently; additional practice is recommended.
>
> **1 — Unskilled** — Is unable to perform the task; additional training required.
>
> ☒ **— Unassigned** — Job sheet task is not required or has not been performed.
>
> ✔ **Evaluator's Note:** Formulate and inform the candidate of the standards for this task (time allowed and number of attempts). Observe the candidate perform the task, check the step/key point under the appropriate attempt number as accomplished, record total time (if appropriate), and then use the rating scale above to assign an overall competency rating. If the candidate is unable to perform any step of this job, have the candidate review the materials and try again.

Introduction | The following job steps provide general procedures for operating air lifting bags. Be thoroughly familiar with the tool, its operating principles, methods, and limitations. *ALWAYS read and follow the manufacturer's directions and cautions before powering or operating a tool.*

!CAUTION: Before powering or operating the following tool, you must be dressed in full protective clothing and wearing eye protection.

Equipment and Personnel |
- Team of at least four firefighters in full protective clothing
- Shims
- Cribbing and shoring blocks and wedges
- Portable air compressor or SCBA or cascade cylinders
- Air lifting bags (medium- or low-pressure)
- Air pressure regulator/regulator hose
- Air bag controller
- Air bag safety hoses
- Adjustable or open-end wrench
- Wrecked automobile(s) or other object(s) appropriate for skill demonstration

Job Steps	Key Points	Attempt No.		
		1	2	3
INSTALLING/INFLATING AIR LIFTING BAGS				
(Command)				
1. Establish command and assign team duties.	1. a. Air bag team	——	——	——
	b. Cribbing team	——	——	——
	c. Air supply team	——	——	——
2. Plan the operation.	2. a. Before starting work	——	——	——
	b. Size(s) of air bag(s) to be used			
	c. Location(s) of air bag(s)	——	——	——
	d. Location(s) of cribbing box(es)	——	——	——
	e. Adequate air supply and sufficient cribbing	——	——	——
(All team members)				
3. Adjust your protective clothing.	3. a. Faceshield lowered	——	——	——
	b. Gloves on	——	——	——
(Cribbing team)				
4. Place a cribbing base for air bag footing.	4. a. On order of command	——	——	——
!CAUTION: Do not reach under the load. A slipped load can cause crushing injury to the hands.	b. Only if recommended by air bag manufacturer	——	——	——
	c. Under solid lifting point (no sharp edges, screws, bolts, or other objects protruding from underside of load)	——	——	——
	d. Solid construction	——	——	——
	!CAUTION: Bags supported by box cribs can be extremely hazardous. Use only solid block formation and then only if the air bag manufacturer specifies that the bag can be supported in this manner.			
	e. Cribbing blocks held by their sides	——	——	——
	f. Top of crib about 1 inch *(25 mm)* from load	——	——	——
	g. Large enough base to provide adequate support			
(Air bag team)				
5. Choose the air bag(s).	5. Correct size or sizes for load to be lifted	——	——	——

Job Steps	Key Points	1	2	3
6. Inspect the air bag(s).	6. a. No air bubbles, bulges, lacerations, or abrasions	—	—	—
	b. Removed from service if damage discovered	—	—	—
(Air supply team) **Perform Steps 7–18 to connect a pressure regulator to the SCBA air cylinder:**				
7. Turn the regulator T-handle.	7. a. Counterclockwise	—	—	—
	b. To NO-PRESSURE position	—	—	—
	c. Until no resistance can be felt	—	—	—
	d. Avoiding unscrewing from body of regulator	—	—	—
8. Turn the regulator output valve.	8. a. Clockwise	—	—	—
	b. Until closed	—	—	—
9. Screw the pressure regulator to the air cylinder.	9. a. To outlet port	—	—	—
	b. Clockwise	—	—	—
10. Tighten the connection.	10. a. *(Handwheel)* Hand-tightened	—	—	—
	b. *(Hex nut)* Tightened with adjustable or open-end wrench	—	—	—
11. Turn the cylinder valve handle.	11. a. To pressurize the regulator	—	—	—
	b. Counterclockwise until fully open	—	—	—
	c. Then clockwise one full turn	—	—	—
12. Observe regulator and cylinder gauges.	12. a. Input gauge same pressure as cylinder gauge	—	—	—
	b. Output gauge indicates zero pressure	—	—	—
13. Turn the regulator handle.	13. a. To set the pressure to be delivered to controller	—	—	—
	b. Clockwise	—	—	—
	c. Until output gauge indicates air bag manufacturer's recommended pressure	—	—	—

Job Steps	Key Points	1	2	3
14. Couple the regulator hose to the controller.	14. To inlet port	—	—	—
15. Pull the hose ferrule (not hose itself).	15. a. Gently	—	—	—
	b. To ensure that connection is secure	—	—	—
16. Turn the regulator output valve.	16. a. To pressurize the system up to the controller	—	—	—
	b. Counterclockwise to fully open position	—	—	—
	c. Clockwise one-half to one full turn	—	—	—
17. Couple the air supply safety hose(s) to the air bag(s).	17. Before bag(s) is (are) positioned	—	—	—
18. Couple the air supply safety hose(s) to the controller.	18. a. To outlet port(s)	—	—	—
	b. Gently pulling hose ferrule(s) (not hose itself) to ensure that connection is secure	—	—	—
Perform Steps 19– 21 to test for proper relief valve operation:				
19. Kink air supply safety hose(s).	19. Between controller and bag(s)	—	—	—
20. Pull the controller lever(s).	20. Toward operator	—	—	—
21. Observe gauges and relief valve(s).	21. Relief valve(s) open at manufacturer's maximum operating pressure	—	—	—
(Air bag team) 22. Position the air bag(s).	22. a. Carried by handles	—	—	—
	b. On solid cribbing base (if recommended by manufacturer) or on solid, level surface containing no puncture hazards	—	—	—
	c. Completely under load	—	—	—
	d. Correct side up as indicated by manufacturer	—	—	—
	e. So that center of bag is directly beneath a smooth, solid portion of load	—	—	—
	f. *(One bag)* So that air supply safety hose faces out	—	—	—

Job Steps	Key Points	Attempt No.		
		1	**2**	**3**
	g. *(Two bags)* So that safety hose to one bag enters from right and hose to other bag enters from left	—	—	—
	h. *(Two bags)* Smaller bag centered on top of larger	—	—	—
	i. Bag(s) sandwiched between cribbing and plywood the size of uninflated bag top (if load presents puncture hazard)	—	—	—
	j. Bag(s) not exposed to materials hotter than 220°F *(104°C)* or temperature extreme listed by manufacturer	—	—	—
23. Position controller and operator.	23. a. Out of danger zone	—	—	—
	b. With unobstructed view	—	—	—
24. Manipulate the controller switches to inflate the air bag(s).	24. a. Pulled toward operator to inflate	—	—	—
	b. Released to stop	—	—	—
	c. Pushed away from operator to deflate	—	—	—
	d. Not overinflating bag(s)	—	—	—
	✔ **Note:** It is not necessary to supply air until the relief valve opens. Bags should be inflated only enough to lift the load and disengage the victim.			
	e. Not underinflating bag(s)	—	—	—
	f. Stopping inflation frequently (to allow cribbing team to increase cribbing or shoring)	—	—	—
	g. Continually monitoring load for any shifting	—	—	—
	h. *(Two bags)* Using following inflation sequence:			
	— Bottom bag inflated first just until top bag firmly contacts load	—	—	—

Job Steps	Key Points	Attempt No. 1	2	3
	— Top bag inflated to maximum pressure	___	___	___
	— Bottom bag inflated further if additional height is needed	___	___	___
(Cribbing team)				
25. Shore the load with cribbing.	25. a. Crosshatch box formation	___	___	___
!CAUTION: Do not reach under the load. A slipped load can cause crushing injury to the hands.	b. As load is being elevated	___	___	___
	c. Holding cribbing and wedging pieces by their sides or handles	___	___	___
	d. Pushing in rear box portions with another piece of cribbing	___	___	___
	e. Wedging to provide maximum contact with load and box crib	___	___	___
	f. Using enough cribbing to support load	___	___	___
	g. Continually monitoring for any shifting	___	___	___
(Air supply team)				
26. Manipulate controller levers to deflate the air bag(s).	26. a. When cribbing is built to desired level	___	___	___
	b. Lever pushed away from operator	___	___	___
	c. Deflating gradually	___	___	___
	d. Until load rests firmly on cribbing	___	___	___
27. Shut off the air supply.	27. a. *(Compressor)* Switched off	___	___	___
	b. *(Cylinder valve)* Turned clockwise until closed	___	___	___
28. Unscrew the handwheel or cylinder nut.	28. a. Slightly	___	___	___
	b. To bleed system at cylinder	___	___	___
29. Turn the pressure regulator output valve.	29. a. Counterclockwise	___	___	___
	b. Until closed	___	___	___
30. Unscrew pressure regulator from cylinder.	30. Counterclockwise	___	___	___
31. Turn the relief valves at the controller.	31. a. Slightly	___	___	___
	b. To bleed system at controller	___	___	___

Job Steps	Key Points	Attempt No. 1 2 3
32. Disconnect the regulator hose from controller.	32. (None)	— — —
33. Disconnect the air supply hose(s).	33. First from controller and then from air bag(s)	— — —
(Air bag team) 34. Remove the airbag(s) from under the load.	34. Monitoring load continually for any shifting	— — —
(All team members) 35. Return equipment to proper storage.	35. Per manufacturer's instructions and department protocol	— — —
	Time (Total)	— — —

Evaluator's Comments _____

_____

_____

CHANGING A COMPRESSED AIR CYLINDER

(Air supply team)

Job Steps	Key Points	Attempt No.
1. Signal the cribbing and air bag teams to stop the lifting operation.	1. Whenever cylinder pressure drops below 200 psi *(1 400 kPa)*	— — —
Perform Steps 2–4 to bleed regulator and controller supply hose(s):		
2. Turn the cylinder valve.	2. a. Clockwise	— — —
	b. Until closed	— — —
3. Unscrew the handwheel or cylinder nut.	3. Slightly	— — —
4. Unscrew the pressure regulator from the cylinder.	4. a. Counterclockwise	— — —
	b. To disconnect regulator from exhausted cylinder	— — —
Perform Steps 5–9 to connect the pressure regulator to the new cylinder:		
5. Turn the regulator output valve.	5. a. Clockwise	— — —
	b. Until closed	— — —

Job Steps	Key Points	1	2	3
6. Screw the pressure regulator to the new air cylinder.	6. a. To outlet port	___	___	___
	b. Clockwise	___	___	___
7. Tighten the connection.	7. a. *(Handwheel)* Hand-tightened	___	___	___
	b. *(Hex nut)* Tightened with adjustable or open-end wrench	___	___	___
8. Turn the cylinder valve.	8. a. Counterclockwise	___	___	___
	b. Slowly	___	___	___
	c. Until open	___	___	___
9. Turn the regulator output valve.	9. a. Counterclockwise	___	___	___
	b. Slowly	___	___	___
	c. Until operating pressure is reached	___	___	___
10. Signal the air bag team to continue the lifting operation.	10. (None)	___	___	___

Time (Total)   ___ ___ ___

Evaluator's Comments _____

_____

_____

Job Sheet 7A-14
Use a Block and Tackle

Name _____ Date _____

Evaluator _____ Overall Competency Rating _____

References	NFPA 1001, Rescue Operations 4-4.2

Essentials, page 195

Prerequisites | None

Student's Instructions | To meet evaluation standards, you must perform this job within _____ *[amount of time, if applicable]*; you may have _____ attempts. When you are ready to perform this job, ask your instructor to observe the procedure and complete this form. To show mastery of this job, you must perform all steps to receive an overall competency rating of at least 2.

Competency Rating Scale

3 — **Skilled** — Meets all evaluation criteria and standards; performs task independently on first attempt; requires no additional practice or training.

2 — **Moderately skilled** — Meets all evaluation criteria and standards; performs task independently; additional practice is recommended.

1 — **Unskilled** — Is unable to perform the task; additional training required.

☒ — **Unassigned** — Job sheet task is not required or has not been performed.

✔ **Evaluator's Note:** Formulate and inform the candidate of the standards for this task (time allowed and number of attempts). Observe the candidate perform the task, check the step/key point under the appropriate attempt number as accomplished, record total time (if appropriate), and then use the rating scale above to assign an overall competency rating. If the candidate is unable to perform any step of this job, have the candidate review the materials and try again.

Introduction | The following job steps provide general procedures for operating a block and tackle. Be thoroughly familiar with the tool, its operating principles, methods, and limitations. *ALWAYS read and follow the manufacturer's directions and cautions before powering or operating a tool.*

!CAUTION: Before operating the following tool, you must be dressed in full protective clothing and wearing eye protection.

Equipment and Personnel |
- Team of at least three firefighters in full protective clothing
- Block and tackle with appropriate length, type, and diameter rope
- Heavy-duty twine
- Wrecked automobile(s) or other heavy object(s) appropriate for skill demonstration

---

Job Steps	Key Points	Attempt No.

1. Adjust your protective clothing.

2. Reeve the block.

1. a. Faceshield lowered   — — —

    b. Gloves on   — — —

2. a. Correct type rope for block and load   — — —

    b. Correct diameter rope for block and load   — — —

    c. Correct length of rope for job   — — —

    d. Following appropriate configuration below   — — —

3. Pull the rope through the block.

4. Hook the block to a support.

5. Mouse or lock the standing block hook.

3. a. To adjust reeving   — — —

    b. Several times   — — —

    c. Rope untwisted   — — —

4. a. Standing block   — — —

    b. High enough for desired lift   — — —

    c. Strong enough to withstand pull and load   — — —

5. *(Latch Locking safety)* Closing safety latch   — — —

    (Mousing)

    a. By wrapping strong twine around hook   — — —

Job Steps	Key Points	Attempt No. 1	2	3
	b. Eight to ten times	___	___	___
	c. Several additional turns around mousing	___	___	___
	d. Ends tied securely	___	___	___
6. Hook the running block to the load.	6. a. At adequate support point	___	___	___
	b. Hook moused or locked	___	___	___
7. Take positions on the fall line.	7. a. Downhill if possible	___	___	___
	b. Everyone on same side of rope	___	___	___
	c. Out of danger if tackle fails or support falls	___	___	___
8. Pull the fall line to lift the load.	8. a. On command of last firefighter on fall line	___	___	___
	b. In direct line with sheaves	___	___	___
	c. Everyone exerting simultaneous and steady pull	___	___	___
	d. Slowly and cautiously	___	___	___
	e. Everyone holding onto gain	___	___	___
	f. Until load is lifted several feet *(meters)* off ground	___	___	___
	g. Gradually easing off pulling	___	___	___
	!CAUTION: If you stop pulling suddenly, the load may swing dangerously.			
9. Secure the fall line to suspend the load.	9. a. Last firefighter on line	___	___	___
	b. While team maintains tension	___	___	___
	c. With free end of fall line	___	___	___
	d. To appropriate stationary object	___	___	___
	e. With clove hitch and overhand safety	___	___	___

Job Steps	Key Points	Attempt No. 1	2	3
10. Return to positions on the fall line.	10. To lower load	___	___	___
11. Pull the fall line.	11. a. On command of last fire-fighter on rope	___	___	___
	b. Unanimously	___	___	___
	c. Just enough to lift load and relieve tension from end of fall line	___	___	___
12. Untie the fall line.	12. a. Last firefighter on line	___	___	___
	b. From stationary object	___	___	___
13. Lower the load.	13. a. Slowly	___	___	___
	b. Hand over hand	___	___	___
	c. Until object rests on ground	___	___	___
14. Unreeve the block and tackle.	14. Reversing reeving process	___	___	___
15. Return the block and tackle to proper storage.	15. Per manufacturer's instructions and department protocol	___	___	___
	Time (Total)	___	___	___

Evaluator's Comments _____

_____

_____

Job Sheet 7A-15
Use an Electric or Gasoline-Powered
Circular Saw

Name _____ Date _____

Evaluator _____ Overall Competency Rating _____

References	NFPA 1001, Rescue Operations 4-4.2
Prerequisites	None
Student's Instructions	To meet evaluation standards, you must perform this job within _____ *[amount of time, if applicable]*; you may have _____ attempts. When you are ready to perform this job, ask your instructor to observe the procedure and complete this form. To show mastery of this job, you must perform all steps to receive an overall competency rating of at least 2.

> **Competency Rating Scale**
>
> **3 — Skilled** — Meets all evaluation criteria and standards; performs task independently on first attempt; requires no additional practice or training.
>
> **2 — Moderately skilled** — Meets all evaluation criteria and standards; performs task independently; additional practice is recommended.
>
> **1 — Unskilled** — Is unable to perform the task; additional training required.
>
> ☒ **—Unassigned** — Job sheet task is not required or has not been performed.
>
> ✔ **Evaluator's Note:** Formulate and inform the candidate of the standards for this task (time allowed and number of attempts). Observe the candidate perform the task, check the step/key point under the appropriate attempt number as accomplished, record total time (if appropriate), and then use the rating scale above to assign an overall competency rating. If the candidate is unable to perform any step of this job, have the candidate review the materials and try again.

Introduction	The following job steps provide general procedures for operating an electric or gasoline-powered circular saw. Be thoroughly familiar with the tool, its operating principles, methods, and limitations. *ALWAYS read and follow the manufacturer's directions and cautions before powering or operating a tool.*
	!CAUTION: Before powering or operating the following tool, you ***must*** be dressed in full protective clothing and wearing safety goggles under your lowered faceshield.
Equipment and Personnel	• One firefighter in full protective clothing and helmet with faceshield • Safety goggles • Charged booster hose • Electric or gasoline-powered circular saw • Fuel for gasoline-powered saw • Manufacturer's recommended oil • Electric power source/outlet • Wrecked automobile(s) or other object(s) appropriate for skill demonstration

Job Steps	Key Points	Attempt No.
		1 2 3
1. Check the area in which the work is to be done. !CAUTION: Contact between the blade and metal may produce sparks that will ignite flammable vapors.	1. a. No ignition hazards b. No flammable vapors	___ ___ ___ ___ ___ ___
2. Check the saw blade.	2. a. Correct for job b. Sharp c. Not using if damaged in any way	___ ___ ___ ___ ___ ___ ___ ___ ___
3. Check the oil level.	3. a. Following manufacturer's instructions b. Adding proper weight oil if necessary	 ___ ___ ___ ___ ___ ___
4. Check saw's safety features.	4. Blade and hand guards in place	___ ___ ___
5. Cool the saw blade.	5. a. When using on metals b. Before cutting c. With a fine water mist from booster line	___ ___ ___ ___ ___ ___ ___ ___ ___
6. Adjust your protective clothing.	6. a. Safety goggles on b. Faceshield lowered c. Gloves on	___ ___ ___ ___ ___ ___ ___ ___ ___
7. Provide power to the saw.	7. a. *(Electric)* Grounded power source b. *(Gasoline)* Proper octane gasoline	 ___ ___ ___ ___ ___ ___
8. Start saw blade rotation.	8. a. Blade not contacting work b. *(Electric)* Switch in ON position c. *(Gasoline)* Compression button depressed (if applicable) and cord pulled d. Per manufacturer's instructions and cautions	___ ___ ___ ___ ___ ___ ___ ___ ___ ___ ___ ___

Job Steps	Key Points	Attempt No.

9. Make the cut(s).

   !CAUTION: Stay alert and keep hands on saw handles and away from material being cut. Power saws can easily sever fingers and limbs.

9. a. Blade perpendicular to work   ___ ___ ___

   b. *(Electric)* Power cord behind operator and away from blade   ___ ___ ___

   c. Not forcing blade through material (blade doing the work)   ___ ___ ___

   d. Cooling blade often with fine water mist (if cutting metal)   ___ ___ ___

10. Stop the saw blade.

10. a. When cut has been made   ___ ___ ___

    b. Releasing trigger   ___ ___ ___

11. *(Electric)* Disconnect the saw from its power source.

11. Pulling plug, not cord   ___ ___ ___

12. *(Gasoline)* Shut off the saw engine.

12. Pushing KILL switch   ___ ___ ___

13. Return the tool to proper storage.

13. Per manufacturer's instructions and department protocol   ___ ___ ___

Time (Total)	___ ___ ___

Evaluator's Comments _____

_____

_____

Job Sheet 7A-16
Use an Electric Reciprocating Saw

Name _____ Date _____

Evaluator _____ Overall Competency Rating _____

References	NFPA 1001, Rescue Operations 4-4.2
Prerequisites	None
Student's Instructions	To meet evaluation standards, you must perform this job within _____ *[amount of time, if applicable]*; you may have _____ attempts. When you are ready to perform this job, ask your instructor to observe the procedure and complete this form. To show mastery of this job, you must perform all steps to receive an overall competency rating of at least 2.

> **Competency Rating Scale**
>
> **3 — Skilled** — Meets all evaluation criteria and standards; performs task independently on first attempt; requires no additional practice or training.
>
> **2 — Moderately skilled** — Meets all evaluation criteria and standards; performs task independently; additional practice is recommended.
>
> **1 — Unskilled** — Is unable to perform the task; additional training required.
>
> ☒ **— Unassigned** — Job sheet task is not required or has not been performed.
>
> ✔ **Evaluator's Note:** Formulate and inform the candidate of the standards for this task (time allowed and number of attempts). Observe the candidate perform the task, check the step/key point under the appropriate attempt number as accomplished, record total time (if appropriate), and then use the rating scale above to assign an overall competency rating. If the candidate is unable to perform any step of this job, have the candidate review the materials and try again.

Introduction	The following job steps provide general procedures for operating an electric reciprocating saw. Be thoroughly familiar with the tool, its operating principles, methods, and limitations. *ALWAYS read and follow the manufacturer's directions and cautions before powering or operating a tool.*
	!CAUTION: Before powering or operating the following tool, you ***must*** be dressed in full protective clothing and wearing eye protection.
Equipment and Personnel	• One firefighter in full protective clothing • Safety goggles • Charged booster hose • Electric reciprocating saw • Electric power source/outlet • Wrecked automobile(s) or other object(s) appropriate for skill demonstration

Job Steps	Key Points	Attempt No.
		1 2 3

1. Check the area in which work is to be done.

   !CAUTION: Contact between the blade and metal may produce sparks that will ignite flammable vapors.

   1. a. No ignition hazards ___ ___ ___
      b. No flammable vapors ___ ___ ___

2. Check the saw blade.

   2. a. Correct for job ___ ___ ___
      b. Sharp ___ ___ ___
      c. Not using if damaged in any way ___ ___ ___

3. Check saw's safety features.

   3. a. Footplate in place ___ ___ ___
      b. Power cord in good condition ___ ___ ___

4. Adjust your protective clothing.

   4. a. Safety goggles on ___ ___ ___
      b. Faceshield lowered ___ ___ ___
      c. Gloves on ___ ___ ___

5. Connect the saw to its power source.

   5. Grounded ___ ___ ___

6. Pick up and position the saw.

   6. a. One hand on handle grip ___ ___ ___
      b. Other hand under saw barrel ___ ___ ___
      c. Power cord behind you and away from blade ___ ___ ___

7. Depress the handle-grip starter button to start the saw blade.

   !CAUTION: Stay alert and keep hands on saw handles and away from material being cut. Power saws can easily sever fingers and limbs.

   7. Blade not contacting work ___ ___ ___

8. Make the cut(s).

   8. a. Blade perpendicular to work ___ ___ ___
      b. Not forcing blade through material (blade doing the work) ___ ___ ___

9. Release the handle-grip starter button to shut off saw blade.

   9. When cut has been made ___ ___ ___

Job Steps	Key Points	Attempt No.		
		1	2	3
10. Remove the saw blade from the work.	10. Without bending blade	—	—	—
11. Disconnect the saw from its power source.	11. Pulling plug, not cord	—	—	—
12. Return the tool to proper storage.	12. Per manufacturer's instructions and department protocol	—	—	—
	Time (Total)	—	—	—

Evaluator's Comments _____

_____

_____

Job Sheet 7A-17
Use an Electric or Gasoline-Powered Chain Saw

Name _____ Date _____

Evaluator _____ Overall Competency Rating _____

References	NFPA 1001, Rescue Operations 4-4.2
Prerequisites	None
Student's Instructions	To meet evaluation standards, you must perform this job within _____ *[amount of time, if applicable]*; you may have _____ attempts. When you are ready to perform this job, ask your instructor to observe the procedure and complete this form. To show mastery of this job, you must perform all steps to receive an overall competency rating of at least 2.

Competency Rating Scale

3 — Skilled — Meets all evaluation criteria and standards; performs task independently on first attempt; requires no additional practice or training.

2 — Moderately skilled — Meets all evaluation criteria and standards; performs task independently; additional practice is recommended.

1 — Unskilled — Is unable to perform the task; additional training required.

☒ **— Unassigned** — Job sheet task is not required or has not been performed.

✔ **Evaluator's Note:** Formulate and inform the candidate of the standards for this task (time allowed and number of attempts). Observe the candidate perform the task, check the step/key point under the appropriate attempt number as accomplished, record total time (if appropriate), and then use the rating scale above to assign an overall competency rating. If the candidate is unable to perform any step of this job, have the candidate review the materials and try again.

Introduction	The following job steps provide general procedures for operating an electric or gasoline-powered chain saw. Be thoroughly familiar with the tool, its operating principles, methods, and limitations. *ALWAYS read and follow the manufacturer's directions and cautions before powering or operating a tool.* **!CAUTION:** Before powering or operating the following tool, you ***must*** be dressed in full protective clothing and wearing eye protection.
Equipment and Personnel	• One firefighter in full protective clothing • Safety goggles • Charged booster hose • Electric or gasoline-powered chain saw • Fuel for gasoline-powered saw • Manufacturer's recommended oil • Electric power source/outlet • Wrecked automobile(s) or other object(s) appropriate for skill demonstration

---

Job Steps	Key Points	Attempt No. 1 2 3
1. Check the area in which work is to be done. !CAUTION: Contact between the blade and metal may produce sparks that will ignite flammable vapors.	1. a. No ignition hazards b. No flammable vapors	___ ___ ___ ___ ___ ___
2. Check the chain teeth.	2. a. Correct for job b. Sharp c. No using if damaged in any way	___ ___ ___ ___ ___ ___ ___ ___ ___
3. Check the chain.	3. a. Oiling if necessary b. Tightening if loose c. Not using if damaged in any way	___ ___ ___ ___ ___ ___ ___ ___ ___
4. Check the oil level.	4. a. Following manufacturer's instructions b. Adding proper weight oil if necessary	___ ___ ___ ___ ___ ___
5. Check saw's safety features.	5. a. Chain and hand guards in place b. Chain brake functional (if applicable)	___ ___ ___ ___ ___ ___
6. Adjust your protective clothing.	6. a. Safety goggles on b. Faceshield lowered c. Gloves on	___ ___ ___ ___ ___ ___ ___ ___ ___
7. Provide power to saw.	7. a. (Electric) Grounded power source b. (Gasoline) Proper octane gasoline	___ ___ ___ ___ ___ ___
8. Start the chain rotation.	8. a. Chain not contacting work b. (Electric) Trigger in ON position c. (Gasoline) Compression button depressed (if applicable) and cord pulled per manufacturer's instructions and cautions	___ ___ ___ ___ ___ ___ ___ ___ ___

Job Steps	Key Points	Attempt No.		
9. Make the cut(s). **!CAUTION:** Stay alert and keep hands on saw handles and away from material being cut. Power saws can easily sever fingers and limbs.	9. a. Chain at effective angle to work	—	—	—
	b. *(Electric)* Power cord behind operator and away from blade	—	—	—
	c. Not forcing blade through material (blade doing the work)	—	—	—
10. Oil chain.	10. a. Often	—	—	—
	b. Per manufacturer's directions	—	—	—
11. Remove chain from the work.	11. When cut has been made	—	—	—
12. Stop the chain rotation.	12. Trigger released	—	—	—
13. Set the saw down.	13. a. When chain has stopped completely	—	—	—
	b. Where it will not be a tripping hazard	—	—	—
14. *(Electric)* Disconnect the saw from its power source.	14. Pulling plug, not cord	—	—	—
15. *(Gasoline)* Shut off the saw engine.	15. Pushing KILL switch	—	—	—
16. Return the tool to proper storage.	16. Per manufacturer's instructions and department protocol	—	—	—
	Time (Total)	—	—	—

Evaluator's Comments _____

_____

_____

STUDENT APPLICATIONS

FOURTH EDITION

ESSENTIALS OF FIRE FIGHTING

LESSON 7B

VEHICLE EXTRICATION & SPECIAL RESCUE

FIREFIGHTER II

FIRE PROTECTION PUBLICATIONS
OKLAHOMA STATE UNIVERSITY

Study Objectives

LESSON OBJECTIVE

After completing this lesson, you will be able to assist a rescue operation team, work as a member of a team to extricate a victim trapped in a motor vehicle, and perform special rescue operations.

ENABLING OBJECTIVES

After reading Chapter 7 of *Essentials,* pages 197 through 214, and completing related activities, you will be able to —

1. List considerations to be made when sizing up a vehicle accident.

2. List concerns of rescuers who assess the situation at automobile accidents.

3. State the purpose of vehicle stabilization.

4. List methods of gaining access to victims in vehicles.

5. List complications of extrication efforts as a result of passenger restraint and protection systems.

6. Select facts about disentanglement and patient management.

7. State the purpose of packaging.

8. Distinguish between *laminated glass* and *tempered glass.*

9. Select the correct method for removing vehicle glass.

10. **Remove automotive window glass. *(Job Sheet 7B-1)***

11. Match vehicle roof posts to their letter designations.

12. **Remove vehicle doors. *(Job Sheet 7B-2)***

13. **Move or remove vehicle roofs. *(Job Sheet 7B-3)***

14. **Remove steering wheels and columns. *(Job Sheet 7B-4)***

15. **Displace dashboards. *(Job Sheet 7B-5)***

16. Match types of building collapse to their descriptions.

17. List the two types of hazards associated with structural collapse rescue operations.

18. Distinguish between *shoring* and *tunneling.*

19. Select facts about trench rescue operations.

20. State the role of fire departments in cave and tunnel rescue operations.

21. Select facts about rescue operations involving electricity.

22. Distinguish between *rescues* and *recoveries.*

23. Describe the methods for performing a water rescue.

24. Describe the methods for performing an ice rescue.

25. List factors that should be taken into account during industrial extrications.

26. Select facts about elevator and escalator rescues.

27. **Assist rescue teams.** *(Practical Activity Sheet 7B-1)*

Study Sheet

Introduction

This study sheet is intended to help you learn the material in Chapter 7 of *Essentials of Fire Fighting*, Fourth Edition, pages 197 through 214. You may use it for self-study, or you may use it to review material that will be covered in the lesson and chapter review tests. The numbers in parentheses are the pages in *Essentials* on which the answers or terms can be found.

Chapter Vocabulary

Be sure that you know the chapter-related meanings of the following terms and abbreviations. Use a dictionary or the glossary in *Fire Service Orientation and Terminology* if you cannot determine the meaning of the term from its context.

- Cribbing *(200)*
- Entrapment *(197)*
- Extrication *(197)*
- Laminated glass *(202)*
- Packaging *(202)*
- Rescue *vs.* recovery *(211)*
- Shoring *(208)*
- Side-Impact Protection System *(201)*
- Size-up *(197)*
- Stabilize *(198)*
- Supplemental Restraint System *(201)*
- Tempered glass *(202)*
- Triage *(198)*
- Tunneling *(208)*
- Types of building collapse *(206)*

Study Questions & Activities

1. What aspects of a vehicle accident should be considered as rescue personnel approach and arrive at the scene? *(197)*

   a. _____

   _____

   b. _____

   _____

   c. _____

   _____

   d. _____

   _____

e. _____

_____

f. _____

_____

g. _____

_____

2. List five conditions that a rescuer should check for when doing a vehicle survey at an accident scene. *(197)*

a. _____

_____

b. _____

_____

c. _____

_____

d. _____

_____

e. _____

_____

3. Ideally, how many vehicles should an individual rescuer survey at an accident scene? *(197)*

_____

4. While rescuers are checking the vehicles, what task should another rescuer perform? *(197, 198)*

_____

5. What task should be performed after assessing the scene if there are entrapped victims? *(198)*

_____

_____

6. What is the primary goal of vehicle stabilization? *(198)*

_____

_____

7. What is the most common method of performing horizontal stabilization? *(199)*

_____

8. What is the recommended minimum number of air lifting bags and their placement for stabilizing a vehicle? *(199)*

_____

_____

9. How should a rescuer hold cribbing blocks when placing them under or removing them from under a vehicle? *(197, 198)*

_____

10. What is a *step block?* *(200)*

_____

11. What are three general methods of gaining access to victims trapped as the result of an automobile accident? *(200)*

    a. _____

    b. _____

    c. _____

12. What danger is presented by vehicle supplemental restraint systems such as air bags? *(201)*

_____

_____

_____

_____

_____

_____

13. What is the purpose of placing a rescuer in the vehicle with victims as disentanglement procedures are in progress? *(202)*

_____

_____

_____

_____

14. What is *packaging* in terms of removing a patient from a vehicle? *(202)*

_____

_____

15. Briefly describe the differences between safety glass and tempered glass. *(202, 203)*

_____

_____

_____

_____

16. List two ways of controlling tempered glass when it is broken. *(204)*

    a. _____

    _____

    b. _____

    _____

17. How are the door posts of a car labeled? *(205)*

    a. Front door hinge post _____

    _____

    b. Rear door hinge post _____

    _____

    c. Rear door latch post _____

    _____

18. What is the best method for extricating a person who is pinned by the dashboard? *(205)*

    _____

    _____

19. Describe the four types of structural collapse. *(206, 207)*

    a. Lean-to _____

    _____

    b. Pancake _____

    _____

    c. V-shape _____

    _____

    d. Cantilever _____

    _____

20. What are the two categories of hazards associated with structural collapse? *(207)*

    a. _____

    b. _____

21. Distinguish between *shoring* and *tunneling*. *(208)*

    a. Shoring _____

    _____

    b. Tunneling _____

    _____

22. List five safety precautions that should be observed by rescuers at a trench rescue. *(208, 209)*

    a. _____

    _____

    b. _____

    _____

    c. _____

    _____

    d. _____

    _____

e. _____
_____

23. What is the general rule for the involvement of fire departments in cave rescue operations? *(210)*

_____
_____
_____
_____

24. What are the three general guidelines for rescues involving electricity? *(210)*

a. _____
_____

b. _____
_____

c. _____
_____

25. What happens to voltage around a downed power line as the distance from the point of contact increases? *(210, 211)*

_____
_____

26. How far should rescuers remain from downed power lines? *(211)*

_____

27. What is the difference between *rescue* and *recovery?* *(211)*

_____
_____
_____
_____

28. Describe the following methods of water rescue? *(212)*

a. REACH _____
b. THROW _____
c. ROW _____
d. GO _____

29. Describe the following methods of ice rescue? *(212)*

a. REACH _____
b. THROW _____
c. GO _____

30. What aspects of an industrial extrication should rescue personnel take into account? *(213)*

a. _____

_____

b. _____

_____

c. _____

_____

d. _____

_____

e. _____

_____

31. What is the most important resource for rescuers to obtain during elevator and escalator rescues? *(213, 214)*

_____

32. What are the two main reasons for communicating with passengers trapped on elevators? *(214)*

a. _____

_____

b. _____

_____

Practical Activity Sheet 7B-1
Assist Rescue Teams

Name _____ Date _____

Evaluator _____ Overall Competency Rating _____

References NFPA 1001, Rescue Operations 4-4.2b
Essentials, pages 197–214

Prerequisites None

Introduction Many types of emergencies require the rescue of accident victims. While these situations may involve providing medical treatment to the accident victim, the most difficult part of the operation often entails those procedures required to make it safe for rescuers to attempt the rescue and protection of the victims during the rescue efforts. Although an emergency situation may seem desperate, proper planning of the rescue operation is mandatory in order to keep the incident from escalating, with would-be rescuers trapped or endangered. This activity sheet will provide you with practice in planning rescue operations for specific situations.

Directions Read the situations given on the following pages. Then answer the questions that accompany each situation.

✔Note: It is not necessary to provide details of medical treatment. The main objective is to describe extrication activities related to the incident.

Activity SCENARIO 1

You have responded to an accident involving three vehicles. Vehicle 1 has overturned, and two people are trapped inside, unconscious and possibly dead. Vehicle 2 is a pickup truck that is lying on its side, leaning against Vehicle 1. The driver of Vehicle 2 is sitting beside his pickup with blood on his head and a great deal of pain in his left leg. Vehicle 3 is approximately 30 feet *(10 m)* from the other two vehicles. Its front end is smashed with the hood pressed up against the windshield. There are apparently two people in the front seat behind deployed air bags.

Questions

1. What is your first priority upon arriving at the scene?

    _____

    _____

    _____

    _____

2. If six rescuers are available, how should rescue personnel be deployed upon arriving?

_____

_____

_____

_____

3. A quick check of the driver of Vehicle 2 shows that he has a gash on his right temple and a possible concussion and that his left leg is broken. How should he be treated? When should he be treated relative to other rescue actions?

_____

_____

_____

_____

_____

_____

_____

_____

4. How should each vehicle be stabilized?

   a. Vehicle 1 _____

   _____

   b. Vehicle 2 _____

   _____

   c. Vehicle 3 _____

   _____

5. A closer examination of Vehicle 1 reveals that the driver is dead and that the front-seat passenger is alive but seriously injured, suffering from a head injury, possible neck and spinal injury, probable broken ribs, and the possibility of internal injuries. The doors cannot be opened, and the only access appears to be through the rear window. What should be done for each victim in order to remove them from the vehicle?

   a. Passenger _____

   _____

   _____

   _____

   b. Driver _____

   _____

   _____

   _____

6. The survey of Vehicle 3 shows that the doors are locked. The doors are somewhat damaged along the A post, but may be operable. The driver and passenger are unresponsive but show no signs of injuries in their upper bodies. The steering wheel is collapsed, and it appears that the victims may be pinned or partially pinned by the dashboard. How should the rescuers proceed in gaining entry to the victims and extricating them?

_____

_____

_____

_____

_____

_____

SCENARIO 2

You are in command of a rescue team responding to a trench cave-in. Witnesses state that three workers have been buried at one end of the trench. One witness says that he suspects that the cave-in was caused by a ruptured gas pipeline.

Questions

1. What is your first priority upon arriving at the scene?

_____

_____

_____

_____

2. What qualifications and equipment are required of the rescue personnel who enter the trench?

_____

_____

_____

_____

3. How should the incident area be prepared for the rescue? Address issues such as preventing further cave-ins and protecting the rescue team.

_____

_____

_____

_____

4. How should the rescuers look for the buried victims?

_____

_____

_____

_____

SCENARIO 3

Your unit arrives at the scene of a vehicle accident. The driver of a car appears to have lost consciousness, lost control of her vehicle, and run into a utility pole. The pole has broken and fallen onto the car, smashing part of the roof. The driver is still in the vehicle and unconscious. Power lines are draped over the vehicle and along the ground on each side of the vehicle.

Questions

1. What are your first priorities upon arriving at the scene?

_____

_____

_____

_____

2. Where should personnel be stationed at the accident scene?

_____

_____

_____

_____

3. What stabilization actions are required at the scene once it is safe to begin rescue operations?

_____

_____

_____

_____

_____

_____

4. The damage to the roof has made it impossible to open the driver's door and prevents removal of the victim through any of the window openings. How should the victim be removed?

_____

_____

_____

_____

_____

_____

Competency Rating Scale

3 — Skilled — Answered questions appropriately per responses suggested in *Instructor's Guide* answers: All 17 criteria check "yes."

2 — Moderately skilled — Answered questions appropriately per responses suggested in *Instructor's Guide* answers: At least 4 criteria in Scenario 1 and 3 criteria in Scenarios 2 and 3 check "yes."

1 — Unskilled — Answers inappropriate per responses suggested in *Instructor's Guide* answers: Fewer than 4 criteria in Scenario 1 and 3 criteria in Scenarios 2 and 3 check "yes".

☒ —Unassigned — Task is not required or has not been performed.

✔ **Evaluator's Note:** Score the product as indicated below. Use the rating scale above to assign an overall competency rating. Record the overall competency rating on both the student's practical activity sheet and competency profile.

To show competency in this objective, the student must achieve an overall rating of at least 2.

Criteria	Yes	No
SCENARIO 1		
1. Stated first priority.	☐	☐
2. Explained how personnel should be deployed upon arriving.	☐	☐
3. Stated appropriate treatment and correct order of treatment.	☐	☐
4. Identified correct procedure for stabilizing vehicles:		
a. Vehicle 1	☐	☐
b. Vehicle 2	☐	☐
c. Vehicle 3	☐	☐
5. Suggested appropriate procedures for removing victims from Vehicle 1:		
a. Passenger	☐	☐
b. Driver	☐	☐
6. Suggested appropriate procedures for removing victims from Vehicle 3.	☐	☐
SCENARIO 2		
1. Stated first priority.	☐	☐
2. Listed appropriate qualifications and equipment.	☐	☐
3. Explained correct procedure for preparing the area for rescue.	☐	☐
4. Stated appropriate search procedure.	☐	☐

Criteria	Yes	No
SCENARIO 3		
1. Stated first priorities.	☐	☐
2. Stationed personnel at accident scene appropriately.	☐	☐
3. Identified appropriate stabilization actions.	☐	☐
4. Suggested appropriate procedure for removing victim.	☐	☐

Job Sheet 7B-1
Remove Automotive Window Glass

Name _____ Date _____

Evaluator _____ Overall Competency Rating _____

References	NFPA 1001, Rescue Operations 4-4.1b ***Essentials,*** page 205
Prerequisites	JS 7A-10 — Use a Pneumatic Chisel/Hammer JS 7A-16 — Use an Electric Reciprocating Saw
Student's Instructions	To meet evaluation standards, you must perform this job within _____ *[amount of time, if applicable];* you may have _____ attempts. When you are ready to perform this job, ask your instructor to observe the procedure and complete this form. To show mastery of this job, you must perform all steps to receive an overall competency rating of at least 2.

Competency Rating Scale

3 — Skilled — Meets all evaluation criteria and standards; performs task independently on first attempt; requires no additional practice or training.

2 — Moderately skilled — Meets all evaluation criteria and standards; performs task independently; additional practice is recommended.

1 — Unskilled — Is unable to perform the task; additional training required.

☒ — Unassigned — Job sheet task is not required or has not been performed.

✔ **Evaluator's Note:** Formulate and inform the candidate of the standards for this task (time allowed and number of attempts). Observe the candidate perform the task, check the step/key point under the appropriate attempt number as accomplished, record total time (if appropriate), and then use the rating scale above to assign an overall competency rating. If the candidate is unable to perform any step of this job, have the candidate review the materials and try again.

Introduction	Frequently rescuers cannot gain access to vehicle accident victims through normal methods and must compromise the vehicle in order to reach them. One avenue of access is to remove window glass. The two common types of window glass — safety or laminated glass and tempered glass — require two different removal techniques. This job sheet covers both of those techniques. ✔Note: A number of tools may be used to remove window glass. This job sheet covers the use of a reciprocating saw, handsaw, or air chisel to remove laminated glass and the use of a spring-loaded center punch to remove tempered glass. The techniques for other tools are essentially the same. Discuss with your instructor the need to practice these procedures with other types of tools.

Equipment and Personnel	• Three to four firefighters (one or two to remove glass and two to protect passengers) in full protective clothing and eye protection • Two passenger dummies • Safety goggles • Tarp, blanket, and backboard or other method of protecting victims • Reciprocating saw, handsaw, or air chisel • Center punch (spring-loaded preferred) • Contact paper, spray adhesive, or other method of controlling broken glass • Equipment as required to stabilize vehicle • Wrecked automobile(s) appropriate for skill demonstration

		Attempt No.		
Job Steps	**Key Points**	**1**	**2**	**3**
PROCEDURE TO PREPARE VEHICLE FOR GLASS REMOVAL *(Command)* 1. Establish command and assign team duties.	1. a. Protection team b. Glass removal team	___ ___	___ ___	___ ___
2. Plan the operation. !CAUTION: The window to be removed should be as far away from the passengers as possible to avoid injuring them.	2. a. Before starting work b. Determining windows to be removed c. Determining method of removing glass	___ ___ ___	___ ___ ___	___ ___ ___
3. Check the area in which the work is to be done. !CAUTION: Operation of power saws can ignite flammable vapors. If flammable vapors are present, use an axe or other method that does not produce ignition sources.	3. a. No ignition hazards b. No flammable vapors	___ ___	___ ___	___ ___
(All team members) 4. Adjust your protective clothing.	4. a. Safety goggles on b. Faceshield lowered c. Gloves on	___ ___ ___	___ ___ ___	___ ___ ___
5. Stabilize the vehicle.	5. a. Stabilizing horizontally with chocks b. Stabilizing vertically if required	___ ___	___ ___	___ ___

Job Steps	Key Points	Attempt No.		
		1	2	3
6. *(Protection team)* Position two firefighters to protect passengers. ✔Note: Assume that access to the interior of the vehicle has been attained.	6. a. One rescuer on either side of passengers	—	—	—
	b. Backboard extended across area between passengers and work area	—	—	—
	c. Tarp or other covering draped over backboard to protect passengers from work area	—	—	—
7. *(Glass-removal team)* Apply contact paper or spray adhesive.	7. a. Per manufacturer's instructions	—	—	—
	b. Covering surface of glass to be removed	—	—	—
	Time (Total)	—	—	—

Evaluator's Comments _____

_____

_____

Job Steps	Key Points	Attempt No.		
		1	2	3
PROCEDURE TO REMOVE LAMINATED GLASS *(Saw operator)*				
1. Cut two slits in the glass to be removed.	1. a. Using reciprocating saw, handsaw, or air chisel	—	—	—
	b. One slit in each upper corner of the window	—	—	—
	c. Per Job Sheet 7A-10 or 7A-16	—	—	—
2. Cut one side of the window.	2. a. Near post	—	—	—
	b. Parallel to post	—	—	—
	c. Length of post	—	—	—
3. Cut the other side of the window.	3. a. Near post	—	—	—
	b. Parallel to post	—	—	—
	c. Length of post	—	—	—
4. Cut the lower portion of the window.	4. Connecting each side cut near the bottom of window	—	—	—
(Saw operator and other glass-removal team member)				
5. Position two firefighters.	5. One on each side of window	—	—	—

Job Steps	Key Points	Attempt No.		
		1	2	3
6. Grasp the glass.	6. a. Each team member	——	——	——
	b. Near bottom cut	——	——	——
7. Raise the glass.	7. a. Moving bottom outward	——	——	——
	b. Using care not to break the glass	——	——	——
8. Remove the glass.	8. a. Pulling down to dislodge from frame	——	——	——
	b. Folding back over roof	——	——	——
9. Dispose of the glass.	9. a. Out of the way of operations	——	——	——
	b. Per local protocol	——	——	——

Time (Total)	——	——	——

Evaluator's Comments _____

_____

_____

PROCEDURE TO REMOVE TEMPERED GLASS		Attempt No.		
(Tool operator)				
1. Place the point of the punch against the window.	1. Lower corner of window	——	——	——
2. Brace hand holding punch to prevent it from going through the window.	2. With other hand	——	——	——
3. Operate the punch.	3. a. Pushing in quickly and releasing	——	——	——
	b. Shattering glass	——	——	——
4. Remove the glass.	4. a. Removing adhering surface with attached glass	——	——	——
	b. Removing glass from edges of window	——	——	——
5. Dispose of the glass.	5. a. Out of the way of operations	——	——	——
	b. Per local protocol	——	——	——

Job Steps	Key Points	Attempt No.		
		1	2	3
	Time (Total)	—	—	—

Evaluator's Comments _____

_____

_____

Job Sheet 7B-2
Remove Vehicle Doors

Name _____ Date _____

Evaluator _____ Overall Competency Rating _____

References	NFPA 1001, Rescue Operations 4-4.1b *Essentials,* page 205
Prerequisites	Job Sheet 7A-4 — Use Hydraulic Spreaders
Student's Instructions	To meet evaluation standards, you must perform this job within _____ *[amount of time, if applicable];* you may have _____ attempts. When you are ready to perform this job, ask your instructor to observe the procedure and complete this form. To show mastery of this job, you must perform all steps to receive an overall competency rating of at least 2.

Competency Rating Scale

3 — Skilled — Meets all evaluation criteria and standards; performs task independently on first attempt; requires no additional practice or training.

2 — Moderately skilled — Meets all evaluation criteria and standards; performs task independently; additional practice is recommended.

1 — Unskilled — Is unable to perform the task; additional training required.

☒ **— Unassigned** — Job sheet task is not required or has not been performed.

✔ **Evaluator's Note:** Formulate and inform the candidate of the standards for this task (time allowed and number of attempts). Observe the candidate perform the task, check the step/key point under the appropriate attempt number as accomplished, record total time (if appropriate), and then use the rating scale above to assign an overall competency rating. If the candidate is unable to perform any step of this job, have the candidate review the materials and try again.

Introduction	During a collision or rollover, vehicle doors often become deformed or jammed to the point that they will not operate normally. When this happens, it is necessary to use hydraulic spreaders to remove the door from its frame.
Equipment and Personnel	• One firefighter in full protective clothing and eye protection (a second firefighter may be required to help protect passengers) • Two passenger dummies • Hydraulic spreaders and power source • Tarp or other protective covering if passengers are accessible • Equipment as required to stabilize vehicle • Wrecked automobile(s) appropriate for skill demonstration

Job Steps	Key Points	Attempt No. 1 2 3
1. Plan the operation.	1. a. Before starting work	___ ___ ___
	b. Method of removing door	___ ___ ___
	c. Impact of related systems (side-impact protection system and electrical components)	___ ___ ___
2. Adjust your protective clothing.	2. a. Faceshield lowered	___ ___ ___
	b. Gloves on	___ ___ ___
3. Stabilize the vehicle.	3. a. Stabilizing horizontally with chocks	___ ___ ___
	b. Stabilizing vertically if required	___ ___ ___
4. Protect the passengers if they are accessible.	4. a. Placing tarp or other covering between passengers and work area	___ ___ ___
	b. Positioning firefighter to ensure covering stays in place and to monitor passenger during operation	___ ___ ___
5. Isolate the door from other systems if necessary.	5. a. De-energizing side-impact system per manufacturer's instructions	___ ___ ___
	b. Disconnecting battery to isolate electric windows, door locks, speakers, and other electric equipment in doors	___ ___ ___
6. Prepare the area for spreaders operation.	6. a. Ensuring that solid surfaces are available to serve as pressure points for spreaders	___ ___ ___
	b. Removing outer door skin if necessary to allow insertion of spreaders	___ ___ ___
7. Insert the spreaders tips.	7. a. Between door and pillar	___ ___ ___
	b. Aligned square with pressure points	___ ___ ___

Job Steps	Key Points	Attempt No. 1 2 3

Job Steps	Key Points	1	2	3
8. Operate the spreaders.	8. a. Per Job Sheet 7A-4	——	——	——
	b. Until door is separated from hinges	——	——	——
9. Remove the spreaders.	9. a. Closing tips per Job Sheet 7A-4	——	——	——
	b. Withdrawing tool from opening	——	——	——
10. Move the door.	10. To area where it will not endanger others or interfere with operations	——	——	——
	Time (Total)	——	——	——

Evaluator's Comments _____

_____

_____

Job Sheet 7B-3
Move or Remove Vehicle Roofs

Name _____ Date _____

Evaluator _____ Overall Competency Rating _____

References	NFPA 1001, Rescue Operations 4-4.1b *Essentials,* page 205
Prerequisites	Job Sheet 7A-5 — Use Hydraulic Shears Job Sheet 7A-15 — Use an Electric or Gasoline-Powered Circular Saw Job Sheet 7A-16 — Use an Electric Reciprocating Saw Job Sheet 7B-1 — Remove Automotive Window Glass
Student's Instructions	To meet evaluation standards, you must perform this job within _____ *[amount of time, if applicable];* you may have _____ attempts. When you are ready to perform this job, ask your instructor to observe the procedure and complete this form. To show mastery of this job, you must perform all steps to receive an overall competency rating of at least 2.

Competency Rating Scale

3 — Skilled — Meets all evaluation criteria and standards; performs task independently on first attempt; requires no additional practice or training.

2 — Moderately skilled — Meets all evaluation criteria and standards; performs task independently; additional practice is recommended.

1 — Unskilled — Is unable to perform the task; additional training required.

☒ **— Unassigned** — Job sheet task is not required or has not been performed.

✔ **Evaluator's Note:** Formulate and inform the candidate of the standards for this task (time allowed and number of attempts). Observe the candidate perform the task, check the step/key point under the appropriate attempt number as accomplished, record total time (if appropriate), and then use the rating scale above to assign an overall competency rating. If the candidate is unable to perform any step of this job, have the candidate review the materials and try again.

Introduction	In many vehicle accidents, the quickest and safest way of reaching victims is to remove the roof. There are two common methods of removing the vehicle roof. The first is to remove the roof completely, and the second is to cut the front roof supports and fold the roof back. This job sheet covers both methods.
Equipment and Personnel	• Six firefighters (one to operate saw, three to assist with lifting and bending of roof, and two to protect passengers) in full protective clothing and eye protection • Two passenger dummies • Safety goggles • Tarp, blanket, and backboard or other method of protecting victims • Hydraulic shears, reciprocating saw, or circular saw • Equipment as required to stabilize vehicle • Wrecked automobile(s) appropriate for skill demonstration

PROCEDURE TO PREPARE VEHICLE FOR ROOF REMOVAL

(Command)

1. Establish command and assign team duties.

2. Plan the operation.

3. Check the area in which the work is to be done.

   !CAUTION: Contact between a saw blade and metal may produce sparks that will ignite flammable vapors.

(All team members)

4. Adjust your protective clothing.

5. Stabilize the vehicle.

6. *(Protection team)* Position two firefighters to protect front seat passengers.

7. *(Tool operator)* Remove the windshield.

---

1. a. Protection team   ___ ___ ___
   b. Assistance team   ___ ___ ___
   c. Saw operator   ___ ___ ___

2. a. Before starting work   ___ ___ ___
   b. Method of removing roof   ___ ___ ___

3. a. No ignition hazards if using saw   ___ ___ ___
   b. No flammable vapors if using saw   ___ ___ ___

4. a. Safety goggles on   ___ ___ ___
   b. Faceshield lowered   ___ ___ ___
   c. Gloves on   ___ ___ ___

5. a. Stabilizing horizontally with chocks   ___ ___ ___
   b. Stabilizing vertically if required   ___ ___ ___

6. a. One rescuer on either side of passengers   ___ ___ ___
   b. Backboard extended across area between passengers and work area   ___ ___ ___
   c. Tarp or other covering draped over backboard to protect passengers from work area   ___ ___ ___

7. a. Using hydraulic shears, reciprocating saw, or circular saw per Job Sheet 7A-5, 7A-15, or 7A-16   ___ ___ ___
   b. Following procedure outlined in Job Sheet 7B-1   ___ ___ ___

Job Steps	Key Points	Attempt No.

| | **Time (Total)** | — — — |

Evaluator's Comments _____

_____

_____

PROCEDURE TO BEND ROOF

(Tool operator)

1. Cut one A post.

   !CAUTION: Compromising the roof may cause the vehicle body to distort. Beware of shifts in the vehicle, and ensure that stabilization is adequate if such a shift occurs.

1. a. Using hydraulic shears, reciprocating saw, or circular saw per Job Sheet 7A-5, 7A-15, or 7A-16   — — —

   b. As near to dashboard as is practical   — — —

2. Cut the other A post.

2. a. Using hydraulic shears, reciprocating saw, or circular saw per Job Sheet 7A-5, 7A-15, or 7A-16   — — —

   b. As near to dashboard as is practical   — — —

3. Cut one B post.

3. a. Using hydraulic shears, reciprocating saw, or circular saw per Job Sheet 7A-5, 7A-15, or 7A-16   — — —

   b. As near to dashboard as is practical   — — —

4. Cut the other B post.

4. a. Using hydraulic shears, reciprocating saw, or circular saw per Job Sheet 7A-5, 7A-15, or 7A-16   — — —

   b. As near to dashboard as is practical   — — —

(Assistance team)

5. Position two firefighters.

5. a. One on each side of vehicle   — — —

   b. Near C post   — — —

6. Hold bar.

6. a. Positioning bar just forward of C post   — — —

Job Steps	Key Points	Attempt No.
		1 2 3

		1	2	3
	b. Against roof	—	—	—
(Tool operator and other assistance-team member)				
7. Position two firefighters.	7. a. One on each side of vehicle	—	—	—
	b. Near A post	—	—	—
8. Bend roof to expose passenger compartment.	8. Toward rear of vehicle	—	—	—

Time (Total) — — —

Evaluator's Comments _____

_____

_____

PROCEDURE TO REMOVE ROOF

		1	2	3
1. *(Tool operator)* Cut all roof posts. **!CAUTION:** Compromising the roof may cause the vehicle body to distort. Beware of shifts in the vehicle and ensure that stabilization is adequate if such a shift occurs.	1. a. Using hydraulic shears, reciprocating saw, or circular saw per Job Sheet 7A-5, 7A-15, or 7A-16	—	—	—
	b. As near to vehicle body as is practical	—	—	—
(Tool operator and assistance team)				
2. Position four firefighters.	2. a. One near each A post	—	—	—
	b. One near each C post	—	—	—
3. Lift the roof.	3. a. Using legs not back	—	—	—
	b. Avoiding twisting motions	—	—	—
4. Move the roof.	4. To an area in which it will not endanger others or interfere with operations	—	—	—

Time (Total) — — —

Evaluator's Comments _____

_____

Job Sheet 7B-4
Remove Steering Wheels and Columns

Name _____ Date _____

Evaluator _____ Overall Competency Rating _____

References	NFPA 1001, Rescue Operations 4-4.1b ***Essentials,*** page 205
Prerequisites	Job Sheet 7A-4 — Use Hydraulic Spreaders Job Sheet 7A-5 — Use Hydraulic Shears Job Sheet 7A-16 — Use an Electric Reciprocating Saw Job Sheet 7B-1 — Remove Automotive Window Glass Job Sheet 7B-3 — Move or Remove Vehicle Roofs
Student's Instructions	To meet evaluation standards, you must perform this job within _____ *[amount of time, if applicable];* you may have _____ attempts. When you are ready to perform this job, ask your instructor to observe the procedure and complete this form. To show mastery of this job, you must perform all steps to receive an overall competency rating of at least 2.

Competency Rating Scale

3 — **Skilled** — Meets all evaluation criteria and standards; performs task independently on first attempt; requires no additional practice or training.

2 — **Moderately skilled** — Meets all evaluation criteria and standards; performs task independently; additional practice is recommended.

1 — **Unskilled** — Is unable to perform the task; additional training required.

⌧ — **Unassigned** — Job sheet task is not required or has not been performed.

✔ **Evaluator's Note:** Formulate and inform the candidate of the standards for this task (time allowed and number of attempts). Observe the candidate perform the task, check the step/key point under the appropriate attempt number as accomplished, record total time (if appropriate), and then use the rating scale above to assign an overall competency rating. If the candidate is unable to perform any step of this job, have the candidate review the materials and try again.

Introduction	Vehicle accident victims often become entrapped by the steering wheel because it extends so far from the dashboard. There are two methods of providing clearance around the steering wheel. The steering wheel can be removed or the steering column can be collapsed. This job sheet covers both methods.
Equipment and Personnel	• Two firefighters (one to operate equipment and one to monitor equipment and passengers) in full protective clothing and eye protection • Two passenger dummies • Safety goggles • Hydraulic spreaders

- Two steel alloy chains, each 8 feet *(2.5 m)* long
- Hydraulic shears, reciprocating saw, or bolt cutters
- Tarp or other covering to protect passengers
- Equipment as required to stabilize vehicle
- Wrecked automobile(s) appropriate for skill demonstration

Job Steps	Key Points	Attempt No.		
		1	2	3
PROCEDURE TO PREPARE VEHICLE FOR STEERING WHEEL AND COLUMN REMOVAL				
1. Plan the operation.	1. a. Before starting work	___	___	___
	b. Determining method of moving steering wheel	___	___	___
	c. Determining impact of related systems (supplemental restraint system and electrical components)	___	___	___
2. Adjust your protective clothing.	2. a. Safety goggles on	___	___	___
	b. Faceshield lowered	___	___	___
	c. Gloves on	___	___	___
3. Stabilize the vehicle.	3. a. Stabilizing horizontally with chocks	___	___	___
	b. Stabilizing vertically if required	___	___	___
4. Protect the passengers if they are accessible.	4. a. Placing tarp or other covering between passengers and work area	___	___	___
	b. Positioning firefighter to ensure covering stays in place and to monitor passengers during operation	___	___	___
5. Isolate the steering column and wheel from other systems if necessary.	5. a. De-energizing supplemental restraint system per manufacturer's instructions	___	___	___
	b. Disconnecting battery to isolate ignition system, horn, wipers, lights, and other electric equipment connected to steering column	___	___	___

Job Steps	Key Points	Attempt No. 1 2 3

6. Remove the windshield or roof if necessary to gain access to steering column or steering wheel.

6. Per Job Sheet 7B-1 or 7B-3   __ __ __

> **Time (Total)**   __ __ __

Evaluator's Comments _____

_____

_____

PROCEDURE USING HYDRAULIC SHEARS, RECIPROCATING SAW, OR BOLT CUTTERS

1. Check the area in which the work is to be done.

   !CAUTION: Contact between the blade and metal may produce sparks that will ignite flammable vapors.

1. a. No ignition hazards   __ __ __

   b. No flammable vapors   __ __ __

2. Cut the component to be removed.

2. a. Using hydraulic shears, reciprocating saw, or bolt cutters per Job Sheet 7A-5 or 7A-16   __ __ __

   b. Cutting sections of wheel if required   __ __ __

   c. Cutting through steering column   __ __ __

3. Remove the component.

3. a. Using care not to injure passengers   __ __ __

   b. Using proper lifting procedures   __ __ __

4. Move the component.

4. To an area in which it will not endanger others or interfere with operations   __ __ __

> **Time (Total)**   __ __ __

Evaluator's Comments _____

_____

_____

PROCEDURE TO COLLAPSE STEERING WHEEL

!WARNING! This procedure is for rear-wheel-drive vehicles only. *Never* **attempt to collapse the steering wheel on a front-wheel-drive vehicle**.

1. Identify the connection points.

   !CAUTION: Ensure that connection points will hold up under the pressure to be exerted.

2. Position the steering column chain.

3. Position the front end of the chain.

4. Open the spreaders.

5. Rest the spreaders on the hood.

6. Secure ends of chain to spreaders tips.

1. a. On steering wheel ___ ___ ___

   b. On front underside of vehicle ___ ___ ___

2. a. Routing chain across steering wheel to apply equal pressure across steering wheel ___ ___ ___

   b. Positioning free ends of chain near middle of hood ___ ___ ___

3. a. Routing chain around front connecting point ___ ___ ___

   b. Positioning free ends of chain near middle of hood ___ ___ ___

4. To fully open position ___ ___ ___

5. a. Positioned perpendicular to length of vehicle ___ ___ ___

   b. One tip near the free ends of the two chains ___ ___ ___

      ✔**Note:** It may be necessary to place tarps, cribbing, or other materials under the spreaders and chains to produce the best angle of attack for the chains.

6. a. Attaching free ends of steering wheel chain to corresponding tip ___ ___ ___

Job Steps	Key Points	Attempt No. 1	2	3
	b. Attaching free ends of front end chain to corresponding tip	—	—	—
	c. Leaving a small amount of slack	—	—	—
7. Position firefighters to monitor operation.	7. a. One to monitor passenger and movement of steering column	—	—	—
	b. One to monitor operation of spreaders and connection to front end	—	—	—
8. Close the spreaders tips.	8. a. Per Job Sheet 7A-4	—	—	—
	b. Until steering column is adequately collapsed	—	—	—
9. Open the spreaders tips.	9. a. Per Job Sheet 7A-4	—	—	—
	b. Until there is enough slack in chains to allow removal of chains from spreaders tips	—	—	—
10. Disassemble the equipment configuration.	10. a. Removing chains	—	—	—
	b. Disassembling and servicing hydraulic spreaders as required by manufacturer's requirements	—	—	—
	Time (Total)	—	—	—

Evaluator's Comments _____

_____

_____

Job Sheet 7B-5
Displace Dashboards

Name _____ Date _____

Evaluator _____ Overall Competency Rating_____

References	NFPA 1001, Rescue Operations 4-4.1b *Essentials,* pages 205, 206
Prerequisites	Job Sheet 7A-5 — Use Hydraulic Shears Job Sheet 7A-6 — Use a Hydraulic Extension Ram Job Sheet 7A-16 — Use an Electric Reciprocating Saw Job Sheet 7B-1 — Remove Automotive Window Glass Job Sheet 7B-3 — Move or Remove Vehicle Roofs
Student's Instructions	To meet evaluation standards, you must perform this job within _____ *[amount of time, if applicable];* you may have _____ attempts. When you are ready to perform this job, ask your instructor to observe the procedure and complete this form. To show mastery of this job, you must perform all steps to receive an overall competency rating of at least 2.

Competency Rating Scale

3 — **Skilled** — Meets all evaluation criteria and standards; performs task independently on first attempt; requires no additional practice or training.

2 — **Moderately skilled** — Meets all evaluation criteria and standards; performs task independently; additional practice is recommended.

1 — **Unskilled** — Is unable to perform the task; additional training required.

☒ — **Unassigned** — Job sheet task is not required or has not been performed.

✔ **Evaluator's Note:** Formulate and inform the candidate of the standards for this task (time allowed and number of attempts). Observe the candidate perform the task, check the step/key point under the appropriate attempt number as accomplished, record total time (if appropriate), and then use the rating scale above to assign an overall competency rating. If the candidate is unable to perform any step of this job, have the candidate review the materials and try again.

Introduction	Rescuers often find that front-end collisions have pushed the dashboard against the front seats, pinning the passengers. Such victims are often severely injured so that extrication must be done quickly and efficiently. This job sheet covers the most common method of displacing a dashboard.
Equipment and Personnel	• Six firefighters (two to operate equipment, two to assist with lifting and bending of roof, and two to protect passengers) in full protective clothing and eye protection • Two passenger dummies • Safety goggles • Tarp, blanket, and backboard or other method of protecting victims

- Hydraulic shears or reciprocating saw
- Two hydraulic extension rams
- Equipment as required to stabilize vehicle
- Wrecked automobile(s) appropriate for skill demonstration

Job Steps	Key Points	Attempt No. 1	2	3
(Command) 1. Establish command and assign team duties.	1. a. Protection team b. Assistance team c. Equipment operators	— — —	— — —	— — —
2. Plan the operation.	2. a. Before starting work b. Determining method of removing roof c. Determining positioning of equipment	— — —	— — —	— — —
(All team members) 3. Adjust your protective clothing.	3. a. Safety goggles on b. Faceshield lowered c. Gloves on	— — —	— — —	— — —
4. Stabilize the vehicle.	4. a. Stabilizing horizontally with chocks b. Stabilizing vertically if required	— —	— —	— —
5. *(Protection team)* Position two firefighters to protect front seat passengers.	5. a. One rescuer on either side of passengers b. Backboard extended across area between passengers and work area c. Tarp or other covering draped over backboard to protect passengers from work area	— — —	— — —	— — —
6. *(Equipment operators)* Remove windshield.	6. a. Using pneumatic chisel hammer or electric reciprocating saw b. Per Job Sheet 7B-1	— —	— —	— —
7. *(Equipment operators and assistance team)* Move or remove roof.	7. Per Job Sheet 7B-3	—	—	—

(Equipment operators)

8. Make a relief cut in A post.

    8. a. Using hydraulic shears or reciprocating saw per Job Sheet 7A-5 or 7A-16   —   —   —

        b. At base of A post on each side of vehicle   —   —   —

        c. At approximately 45-degree angle into frame or rocker panel   —   —   —

        d. No more than halfway through frame or rocker panel   —   —   —

9. Position the extension rams.

    9. a. One on each side of vehicle   —   —   —

        b. Ram body secured along rocker panel or against bottom of B or C posts   —   —   —

        c. Ram rod positioned against dashboard assembly or top of A posts   —   —   —

10. Operate extension rams.

  10. a. Per Job Sheet 7A-6   —   —   —

        b. Until dashboard is moved clear of passengers   —   —   —

11. Place cribbing or block in the relief cut to hold dashboard in displaced position.

  11. One on each side of vehicle   —   —   —

12. Remove the extension rams.

  12. a. Per Job Sheet 7A-6   —   —   —

        b. Retracting ram enough to relieve pressure on ram   —   —   —

| **Time (Total)** | —   —   — |

Evaluator's Comments _____

_____

_____

Chapter 7 Review Test

→**Directions:** This review test covers the Firefighter II material in Chapter 7 of your *Essentials of Fire Fighting* text. It may be assigned as a study aid (self-test) or may be administered by your instructor as a pretest or posttest.

When used as a study aid, try to answer the questions without referring to the page numbers in *Essentials* or your *Firefighter II Student Applications* workbook *(SA)* on which the answers can be found until after you have completed the entire test. Then check your answers against those on the pages provided in parentheses.

When administered by your instructor as a pretest or posttest, read each of the test questions carefully. Choose the best response and then darken the corresponding letter on your answer sheet.

This chapter review test contains 100 multiple-choice questions, each worth 1 point. To pass the test, you must achieve at least 84 of the 100 points possible.

1. What is the purpose of an *inverter? (186)*

    A. Changing three-phase power to single-phase power

    B. Changing single-phase power to three-phase power

    C. Changing 110- or 220-volt AC current to 12- or 24-volt DC current

    D. Changing 12- or 24-volt DC current to 110- or 220-volt AC current

2. What tool is illustrated below? *(193)*

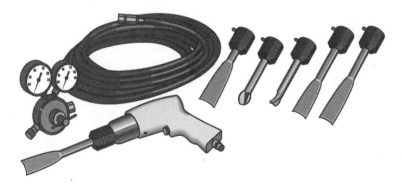

    A. Hydraulic ram   C. Air chisel

    B. Reciprocal saw   D. Hydraulic shears

3. What is the purpose of the tool illustrated in Question 2? *(193)*

    A. Puncturing, cutting, driving

    B. Lifting, displacing, plugging

    C. Separating

    D. Compressing

---

4. Firefighter A says that generators used for emergency services are generally either portable or vehicle-mounted.

Firefighter B says that gasoline-powered generators should not be used at incident sites because of the spark ignition hazard that they present.

Who is right? *(186)*

A. Firefighter A

B. Firefighter B

C. Both A and B

D. Neither A nor B

5. What is the primary purpose of the extrication tool illustrated below? *(189)*

A. Cutting

B. Pushing or pulling

C. Shearing

D. Lifting

6. Which of the following statements is **not** correct? *(186)*

A. Inverters are fuel efficient and produce little noise.

B. Inverters have limited power production capability and limited mobility from the vehicle.

C. Portable generators are not restricted to use at an apparatus, and they produce little noise.

D. Vehicle-mounted generators generally have greater power-production capabilities than portable generators.

7. What are the two general categories of lighting equipment? *(187)*

A. Opaque and transparent

B. Electrical and internal combustion

C. Vertical and rotational

D. Fixed and portable

8. Which of the following is a general guideline for the use of lighting equipment? *(187)*

A. Do not use lighting equipment in wet weather.

B. Match the amount of lighting to the amount of power available and requirements of the situation.

C. Do not use boom lighting unless the lights can be raised above the point of the highest level of fire fighting operations.

D. Match the amount of lighting to the number of personnel and apparatus involved in the response.

9. To obtain a secure, safe connection when setting up emergency power and lighting equipment, what type of electrical receptacles and adapters should be used? *(187)*

    A. Normal

    B. Twist-lock

    C. UL-rated

    D. GFCI

10. Firefighter A says that electrical outlets used at emergency sites should be equipped with ground-fault circuit interrupters.

    Firefighter B says that electrical outlets used at emergency sites do **not** need to conform to *NFPA 70E Standard for Electrical Safety Requirements for Employee Workplaces.*

    Who is right? *(188)*

    A. Firefighter A   C. Both A and B

    B. Firefighter B   D. Neither A nor B

11. Why are electrical adapters carried to responses? *(188)*

    A. To ensure that electrical connections will be UL-rated

    B. To allow resources to exchange equipment in mutual aid situations

    C. To allow electrical equipment of one voltage to be connected to a current source of another value such as 110 voltage to 220 voltage

    D. None of the above; electrical adapters should not be used at emergency sites

12. What is the recommended interval for testing portable power plants? *(188)*

    A. Daily   C. Monthly

    B. Weekly   D. Annually

13. What is the primary reason for wearing gloves when changing quartz light bulbs?*(188)*

    A. The keep-alive element in quartz bulbs will cause them to remain hot even if they have not been recently illuminated.

    B. Oil on the skin will permanently mar the finish of quartz bulbs.

    C. The surface of the bulb is covered with small bits of refractive glass that can become embedded in flesh.

    D. Oil on the skin can cause an energized bulb to explode.

14. To ensure freshness, how often should spare gasoline supplies for portable power equipment be changed? *(188)*

    A. Daily   C. Every other week

    B. Weekly   D. Every three weeks

15. What is the maximum spreading range of hydraulic spreaders?*(189)*

    A. 32 inches *(813 mm)*   C. 8 inches *(200 mm)*

    B. 16 inches *(400 mm)*   D. 4 inches *(100 mm)*

16. Firefighter A says that electrical cable used at emergency sites should be waterproof and explosionproof.

Firefighter B says that electrical cable used at emergency sites should have adequate insulation and no exposed wires.

Who is right? *(187, 188)*

A. Firefighter A

B. Firefighter B

C. Both A and B

D. Neither A nor B

17. What is the approximate opening width of hydraulic shears? *(189)*

A. 28 inches *(710 mm)*

B. 21 inches *(530 mm)*

C. 14 inches *(350 mm)*

D. 7 inches *(180 mm)*

18. Generally, how do the capabilities of hydraulic combination spreader/shears compare to units made specifically for one purpose or the other? *(189)*

A. Combination units are more effective both as spreaders and as shears.

B. Combination units are less effective both as spreaders and as shears.

C. Combination units are more effective as spreaders but less effective as shears.

D. Combination units are less effective as spreaders but more effective as shears.

19. Firefighter A says that hydraulic extension rams are designed for straight pushing operations and cannot be used for pulling operations.

Firefighter B says that hydraulic extension rams can be used for either pushing or pulling operations.

Who is right? *(189)*

A. Firefighter A

B. Firefighter B

C. Both A and B

D. Neither A nor B

20. What is the approximate maximum spread of hydraulic extension rams? *(189)*

A. 36 inches *(900 mm)*

B. 48 inches *(1 200 mm)*

C. 63 inches *(1 600 mm)*

D. 96 inches *(2 400 mm)*

21. Which of the following statements correctly describes the relationship between the opening force and the closing force of hydraulic extension rams? *(189)*

A. The opening force is about one-half that of the closing force.

B. The closing force is about one-half that of the opening force.

C. The opening force is about one-tenth that of the closing force.

D. The closing force is about one-tenth that of the opening force.

22. What type of jack is illustrated below? *(190)*

   A. Bar screw

   B. Hydraulic

   C. Ratchet-lever

   D. Trench screw

23. When raising objects, jacks should always be used in conjunction with ___. *(190)*

   A. Air lifting bags

   B. Cribbing

   C. Shoring

   D. Another jack

24. What type of jack is illustrated below? *(190)*

   A. Bar screw

   B. Hydraulic

   C. Ratchet-lever

   D. Trench screw

25. How often should screw jacks be checked for wear and damage? *(190)*

   A. After every use

   B. Monthly

   C. Annually

   D. Only after being subjected to heavy stress or rough handling

26. Which of the following is ***not*** a correct guideline for the use of jacks? *(190, 191)*

   A. Place jacks on flat, level footing.

   B. If a rescuer must get under a vehicle supported only by jacks, always chock the vehicle wheels.

   C. On soft ground, a board or steel plate with wood on top should be placed under the jack.

   D. Use a ratchet-lever jack only if no other type of jack is available or appropriate for the situation.

27. What type of jack is illustrated below? *(190)*

    A. Bar screw   C. Ratchet-lever

    B. Hydraulic   D. Trench screw

28. Firefighter A says that the primary advantage of the Porta-power® tool system is that it has a complex array of accessories and is very time efficient.

    Firefighter B says that the primary disadvantage of the Porta-power® tool system is that it cannot be operated in narrow places.

    Who is right? *(190)*

    A. Firefighter A   C. Both A and B

    B. Firefighter B   D. Neither A nor B

29. What is the portion of a screw jack that contacts the load called? *(190, 191)*

    A. Base   C. Thread

    B. Footplate   D. Stabilizer

30. What is the primary use of a bar screw jack? *(191)*

    A. Supporting objects to hold them in place

    B. Lifting structural collapses

    C. Raising vehicles involved in accidents

    D. Raising apparatus to stabilize it

31. What is the primary use of a trench screw jack? *(191)*

    A. Displacing dashboards for vehicle extrications

    B. Holding coverings and bridge planks up over trenches

    C. Bracing to dam piles of dirt removed from a trench

    D. Serving as a cross brace for trench rescue operations

32. Firefighter A says that cribbing blocks should be painted or varnished on all surfaces to protect the blocks from moisture and rot.

    Firefighter B says that the end of cribbing blocks should be painted with colors that identify their length.

    Who is right? *(192)*

    A. Firefighter A   C. Both A and B

    B. Firefighter B   D. Neither A nor B

33. Which of the following is a correct guideline for the use of cribbing? *(192)*

    A. Do not drill holes through the cribbing blocks because this weakens the blocks and allows moisture to enter.

    B. Use wedges to shim up loose cribbing formations.

    C. Cribbing that is not quite straight or that has knots and splits is acceptable to use.

    D. Do not use cribbing made of synthetic materials or materials other than wood.

34. Which of the following is *not* an acceptable source of air to power pneumatic tools? *(192)*

    A. Apparatus brake system air compressors

    B. Compressed oxygen supplies

    C. SCBA cylinders or cascade cylinder systems

    D. Vehicle-mounted air compressors

35. A useful tool for assembling wooden structures, such as shoring, is the ___. *(193)*

    A. Pneumatic hammer

    B. Pneumatic nailer

    C. Impact hammer

    D. Air chisel

36. Firefighter A says that sparks from air chisel operations may provide an ignition source for flammable vapors.

    Firefighter B says that air chisels normally operate on air pressures between 100 and 150 psi *(700 kPa and 1 050 kPa)*, though higher pressures may be required for metal thicker than heavy-gauge sheet metal.

    Who is right? *(193)*

    A. Firefighter A

    B. Firefighter B

    C. Both A and B

    D. Neither A nor B

37. How are tripods commonly used to support rescue operations? *(193)*

    A. To brace shoring in trench operations

    B. To create an anchor point over a utility cover or other opening

    C. To support auxiliary lighting equipment

    D. To provide a pivot point for winches when the cable must pass around a corner

38. What should be the relationship between a winch and its load? *(194)*

    A. The winch should be as far from the load as possible.

    B. The winch should be as close to the load as possible.

    C. The winch operator should be as close to the load as possible.

    D. The winch operator should be as close to the winch as possible.

39. The danger zone for winch operations is defined as an area centered on the winch and forming a ___. *(194)*

   A. Circle the radius of which is equal to the length of the deployed cable

   B. Circle the radius of which is equal to half the length of the deployed cable

   C. Semicircle in front of the winch the radius of which is equal to the length of the deployed cable

   D. Semicircle in front of the winch the radius of which is equal to the length of the deployed cable

40. What tool safety violation is illustrated below? *(194)*

   A. Use of a faceshield that may obstruct vision

   B. Incorrect location of winch operator

   C. Use of gloves while operating a remote control device

   D. Use of a remote control device when a control lever is available on the bumper

41. What is the common name of a manually ratcheting tool for pulling and lifting? *(194)*

   A. Winch   C. Cable ratchet

   B. Chain puller   D. Come-along

42. What type of chain is acceptable for rescue operations if it is the correct size? *(194, 195)*

   A. Common chain   C. Proof coil chain

   B. Hardware chain   D. Alloy steel chain

43. How are air lifting bags used in rescue operations classified? *(195)*

   A. Pressure rating   C. Shape

   B. Exterior material   D. Color

44. Firefighter A says that the smaller the pressure rating, the smaller the size of the air lifting bag.

   Firefighter B says high-pressure air lifting bags inflate to a height of 20 inches *(500 mm)* while bags of lower pressures are capable of lifting a load 6 feet *(2 m)*.

   Who is right? *(195)*

   A. Firefighter A   C. Both A and B

   B. Firefighter B   D. Neither A nor B

45. What is the maximum number of air lifting bags that can be stacked to raise a load? *(196)*

    A. 1 (Air lifting bags cannot be stacked.)

    B. 2

    C. 3

    D. 4

46. Air lifting bags should not be exposed to temperatures in excess of ___. *(196)*

    A. 110°F *(43°C)*   C. 330°F *(166°C)*

    B. 220°F *(104°C)*   D. 440°F *(227°C)*

47. What tool safety violation is illustrated below? *(195, 196)*

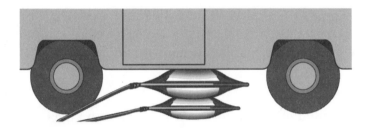

    A. More than one bag is used.

    B. The bags are overinflated.

    C. A larger bag is placed on top of a smaller one.

    D. Separate pneumatic lines are used for the two bags.

48. Which of the following statements is *not* true with regard to the use of block and tackle? *(196, 197)*

    A. Block and tackle assemblies increase the mechanical advantage of a pulling force to provide greater lifting power.

    B. The block contains a frame and one or more pulleys or sheaves.

    C. Tackle refers to the arrangement of the block and ropes.

    D. Block and tackle used for life safety devices should be inspected once per week.

49. What is the maximum length of pipe that can be used with a trench screw jack? *(191)*

    A. 2 feet *(0.6 m)*   C. 6 feet *(1.8 m)*

    B. 4 feet *(1.2 m)*   D. 8 feet *(2.4 m)*

50. Firefighter A says that if air lifting bags are stacked on top of each other, the top bag should be inflated first.

    Firefighter B says that it is preferable to use a single multicell air lifting bag rather than to use stacked bags.

    Who is right? *(196)*

    A. Firefighter A   C. Both A and B

    B. Firefighter B   D. Neither A nor B

51. Ideally, how should rescuers be assigned in assessing the condition of the vehicles involved in an incident? *(197)*

  A. One person should evaluate all involved vehicles to provide a unified assessment.

  B. An individual should be assigned to each vehicle if possible.

  C. First responders should await the arrival of qualified EMS personnel to arrive unless the vehicles are involved in a fire.

  D. Each vehicle should be approached by a team of two rescuers.

52. What factor(s) should be considered when placing the apparatus at the scene of a motor vehicle accident? *(197)*

  A. It should be close enough to the scene so that equipment and supplies are readily available.

  B. It should be far enough from the scene so that it does not interfere with response activities.

  C. It should be positioned to provide a barrier against oncoming traffic.

  D. All of the above are correct.

53. In addition to checking the vehicles, what should a vehicle accident size-up include? *(198)*

  A. Observe those who check the vehicles for safety reasons.

  B. Check a larger area around the vehicles for other vehicles, victims thrown from the vehicles, and other hazards.

  C. Perform triage of the victims and give emergency medical care.

  D. Stabilize the vehicles.

54. Which type of victim should be removed first from vehicles involved in accidents? *(198)*

  A. Those who are not trapped

  B. Those who are trapped

  C. Those with minor injuries who are trapped

  D. Those with serious injuries who are not trapped

55. Firefighter A says that in vehicle extrication, *packaging* is performed before the victim is removed from the vehicle.

  Firefighter B says that *packaging* is performed after the victim is removed from the vehicle to prepare the victim for transportation.

  Who is right? *(202)*

  A. Firefighter A

  B. Firefighter B

  C. Both A and B

  D. Neither A nor B

56. What type of collapse is shown in the illustration below? *(206, 207)*

A. V-shaped

B. Pancake

C. Cantilever

D. Lean-to

57. Which of the following guidelines applies to the use of a vehicle's mechanical system in providing stabilization? *(199)*

A. If operable, the mechanical systems can be used as the sole means of stabilization.

B. The mechanical systems can be used as the sole means of support only if both the parking brake and transmission can be locked (manual transmission in gear and automatic transmission in park).

C. If operable, the mechanical systems can be used to supplement other methods.

D. Whether or not the systems are operable, the vehicle's mechanical systems should not be used.

58. What type of collapse is shown in the illustration below? *(207)*

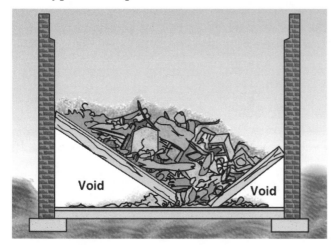

A. Pancake

B. Lean-to

C. Cantilever

D. V-shaped

59. What should be done after the size-up of a vehicle accident is complete and the results have been reported to the incident commander? *(198)*

    A. Rescue the victims.

    B. Stabilize the vehicle(s).

    C. Wash the area down to prevent combustion of spilled fuel.

    D. Remove all window glass to reduce hazards and improve access.

60. What constitutes the minimum appropriate protective equipment for water and ice rescues? *(211)*

    A. Water rescue helmet, an appropriate personal floatation device, and SCBA

    B. Water rescue helmet, an appropriate personal floatation device, and thermal protective suit as required

    C. Turnout clothing and SCBA

    D. Turnout clothing, an appropriate personal floatation device, and SCUBA

61. Which of the following is a last-resort water rescue tactic? *(212)*

    A. Using a boat

    B. Swimming to and dragging victim in

    C. Extending a long pole

    D. Throwing a rope

62. Firefighter A says that even if a vehicle is upright with all wheels on the ground, it should be stabilized to prevent both horizontal and vertical movement.

    Firefighter B says that vehicle wheels should be chocked on the downhill side of the wheels or in both directions of the wheels if the vehicle is on level ground.

    Who is right? *(198, 199)*

    A. Firefighter A   C. Both A and B

    B. Firefighter B   D. Neither A nor B

63. What is the minimum number of air lifting bags needed to prevent vertical movement when stabilizing a vehicle? *(199)*

    A. One   C. Three

    B. Two   D. Four

64. If step blocks are used to stabilize a vehicle, what is the minimum number of blocks that should be used? *(200)*

    A. One   C. Three

    B. Two   D. Four

65. When performing vehicle extrication, what is the first method rescuers should use to gain access to the vehicle? *(200)*

    A. Opening a window   C. Opening a door

    B. Removing the windshield   D. Flapping the roof

66. How far above the trench should exit ladders extend for trench rescue? *(209)*

   A. At least 3 rungs

   B. 3 feet *(1 m)*

   C. At least 14 inches *(350 mm)*

   D. Top rung even with the top of the trench

67. Which of the following should be the first-choice method for ice rescue if the victim is conscious? *(212)*

   A. Get as close to the victim as possible, and use a pole or other device to reach out to the victim.

   B. Crawl out to the victim on ladders.

   C. Break the ice between the shore and the victim to allow the victim to work toward the shore.

   D. Throw a rope to the victim, staying as far as possible from the opening in the ice.

68. When breaking glass to gain access to a vehicle, which window should be broken? *(204)*

   A. Window farthest from trapped victims

   B. Windshield

   C. Window closest to trapped victims

   D. Rear window

69. What type of collapse is shown in the illustration below? *(207)*

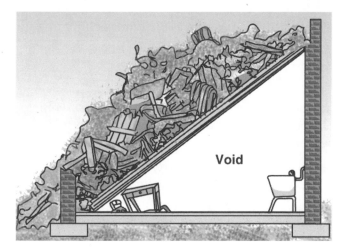

   A. V-shaped   C. Cantilever
   B. Pancake   D. Lean-to

70. Which term designates a vehicle's front post area where the front door is connected to the body? *(205)*

   A. A post   C. F post
   B. W post   D. C post

71. How may a rescuer control glass when breaking a tempered glass window? *(204)*

    A. Contact paper or spray coating

    B. Fly paper or glue

    C. Construction paper or liquid plastic

    D. Wax paper or paste

72. Firefighter A says that there is no danger to rescuers from electrically activated vehicle air bag restraint systems if the vehicle battery has been disconnected or drained.

    Firefighter B says that the best approach for protecting rescuers from injuries from mechanically operated air bags is to break the connection between the sensor and the air bag inflation unit.

    Who is right? *(201)*

    A. Firefighter A   C. Both A and B

    B. Firefighter B   D. Neither A nor B

73. Where is laminated glass most commonly used in vehicles? *(202)*

    A. Rear window   C. Side windows

    B. Windshield   D. All windows

74. Which of the following guidelines applies to the openings made to allow removal of a victim trapped in a vehicle? *(202)*

    A. Sharp edges should be covered.

    B. Edges of openings should be padded.

    C. The opening should be widened enough to allow removal without jerking or sudden movements.

    D. All of the above are correct.

75. Firefighter A says that once access has been gained to a victim trapped in a vehicle, one rescuer should get in the vehicle with the victim to provide medical attention and to protect the victim during disentanglement.

    Firefighter B says that the act of disentanglement involves removing the vehicle from around the victim rather than removing the victim from the vehicle.

    Who is right? *(202)*

    A. Firefighter A

    B. Firefighter B

    C. Both A and B

    D. Neither A nor B

76. What breaking pattern is exhibited by laminated glass? *(202)*

    A. Laminated glass does not break.

    B. It cracks along the edges and pops out as one piece.

    C. It shatters into very small, crumbly pieces.

    D. It breaks in long, pointed shards with sharp edges.

77. What is the composition of laminated glass? *(202)*

   A. Two pieces of plastic with a layer of glass sandwiched between them

   B. Two pieces of glass with a sheet of plastic sandwiched between them

   C. One layer of glass and one layer of plastic bonded together

   D. Three or more layers of glass bonded together

78. Firefighter A says that tempered glass is often used in the side windows of vehicles.
   Firefighter B says that tempered glass is sometimes used in the rear windows of vehicles.

   Who is right? *(203)*

   A. Firefighter A   C. Both A and B

   B. Firefighter B   D. Neither A nor B

79. What breaking pattern is exhibited by tempered glass? *(203)*

   A. It cracks along the edges and pops out as one piece.

   B. It shatters into very small, crumbly pieces.

   C. It breaks in long, pointed shards with sharp edges.

   D. Tempered glass does not break.

80. What is the correct sequence of actions to remove the windshield illustrated below? *(203)*

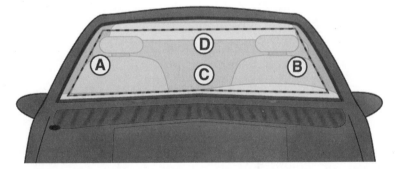

   A. Make cuts along lines A and B and then C and D so the windshield can be pulled out.

   B. Make cuts along lines A and B and then C and D so the windshield can be pushed in.

   C. Make cuts along lines A and B and then C and then fold back onto the roof along line D.

   D. Make cuts along lines A and B and then D and then fold down onto the hood along line C.

81. When using a spring-loaded center punch to break tempered glass, how should the hands be positioned? *(203–204)*

   A. Both hands holding the tool

   B. One hand holding the tool and the other pressing against the glass

   C. One hand holding the tool and the other bracing the tool hand from going into the glass

   D. One hand holding the tool and the other bracing the rescuer's weight against the vehicle

82. What action is recommended when it is necessary to remove the roof from a unibody construction vehicle? *(205)*

    A. Remove the doors first.

    B. Place an extra support, such as a step block, under the rear post of the vehicle.

    C. Place an extra support, such as a step block, under the middle post of the vehicle.

    D. Attach winch cables to the front and rear of the vehicle.

83. Firefighter A says that it is not necessary to remove the windshield and remove or fold back the roof in order to displace the dashboard of a vehicle.

    Firefighter B says that displacing the dashboard requires that cuts be made in the front door posts near the rocker panel so that a hydraulic ram can be placed on each side to push the dashboard up and away from the front seat.

    Who is right? *(205)*

    A. Firefighter A

    B. Firefighter B

    C. Both A and B

    D. Neither A nor B

84. What is the rescue priority in the event of a building collapse? *(206)*

    A. Most seriously injured victims, no matter what their location

    B. Victims in the most heavily trapped locations

    C. Surface and lightly trapped victims

    D. Least seriously injured victims, no matter how heavily tapped

85. Which type of building collapse is *least* likely to contain voids? *(206)*

    A. V-shaped   C. Pancake

    B. Lean-to   D. Cantilever

86. In a building collapse operation, atmospheric contamination and temperature extremes are examples of what type of hazard? *(208)*

    A. Conditional   C. Environmental

    B. Physical   D. Proximity

87. How many rescuers are recommended to perform windshield removal? *(203)*

    A. One to cut and remove the glass

    B. Two to cut and remove the glass

    C. Three: two to cut the glass and one to protect the passengers trapped in the vehicle

    D. Four: two to cut the glass and two to protect the passengers trapped in the vehicle

88. In a building collapse operation, rubble and the potential for secondary collapse are examples of what type of hazard? *(208)*

    A. Conditional   C. Physical

    B. Environmental   D. Proximity

89. Firefighter A says that an air chisel, axe, reciprocating saw, or handsaw can be used to cut laminated glass.

Firefighter B says that it is not necessary to remove the windshield prior to removing the roof.

Who is right? *(203)*

A. Firefighter A

B. Firefighter B

C. Both A and B

D. Neither A nor B

90. What are procedures used to stabilize unstable structures or parts of structures called? *(208)*

A. Tunneling

B. Cribbing

C. Shoring

D. Chocking

91. Which of the following describes the designations of vehicle door posts commonly used by rescuers? *(205)*

A. The front post is A, the rear post is B, and the middle post is not designated.

B. The front post is A, the middle post is B, and the rear post is C.

C. The front post is F, the middle post is M, and the rear post is R.

D. The front post is F, the rear post is R, and the middle post is not designated.

92. Rescuers working in trenches should use SCBA ___. *(209)*

A. Any time they enter a trench

B. If the trench has been found to be oxygen-deficient or contaminated

C. Whenever they enter a trench that is not mechanically ventilated

D. If there are victims to be rescued trapped in the trench

93. What word is defined as *the removal of small debris to create a path to a victim? (208)*

A. Chocking

B. Cribbing

C. Shoring

D. Tunneling

94. Firefighter A says when a rescue involves downed electrical lines, fire department personnel should de-energize the source of power immediately.

Firefighter B says when a rescue involves downed electrical lines, fire department personnel should assume that the lines are energized and control the scene.

Who is right? *(210)*

A. Firefighter A

B. Firefighter B

C. Both A and B

D. Neither A nor B

95. When dealing with downed power lines, personnel should maintain a distance from the lines equal to ___. *(211)*

A. The distance between utility poles

B. The height of one utility pole

C. Half the height of one utility pole

D. Half the distance between utility poles

96. Which type of building collapse occurs in multistory buildings and results in unsupported floors extending from a sidewall? *(207)*

    A. V-shaped

    B. Lean-to

    C. Pancake

    D. Cantilever

97. Which type of building collapse is most likely to result in secondary collapse? *(207)*

    A. Cantilever

    B. V-shaped

    C. Lean-to

    D. Pancake

98. Industrial extrications often require the use of ___. *(213)*

    A. Multijurisdictional responses

    B. Investigations by insurance companies prior to cleanup

    C. Experts such as plant personnel or doctors

    D. Auxiliary lighting

99. What is the preferred approach for elevator rescues? *(214)*

    A. Allow elevator service personnel to respond if lives are not in immediate danger.

    B. Move the elevator to the lowest level before attempting rescue.

    C. Manually maneuver cables to move the elevator to the nearest floor.

    D. Remove all power from the elevator before attempting rescue to reduce electrical hazards.

100. Firefighter A says that rescuers should contact victims of elevator or escalator accidents to reassure them that help is on the way.

    Firefighter B says that escalators should be stopped during rescues or whenever firefighters are advancing hoselines over them.

    Who is right? *(214)*

    A. Firefighter A

    B. Firefighter B

    C. Both A and B

    D. Neither A nor B

REVIEW TEST ANSWER SHEET

	A	B	C	D
1.	○	○	○	○
2.	○	○	○	○
3.	○	○	○	○
4.	○	○	○	○
5.	○	○	○	○
6.	○	○	○	○
7.	○	○	○	○
8.	○	○	○	○
9.	○	○	○	○
10.	○	○	○	○
11.	○	○	○	○
12.	○	○	○	○
13.	○	○	○	○
14.	○	○	○	○
15.	○	○	○	○
16.	○	○	○	○
17.	○	○	○	○
18.	○	○	○	○
19.	○	○	○	○
20.	○	○	○	○
21.	○	○	○	○
22.	○	○	○	○
23.	○	○	○	○
24.	○	○	○	○
25.	○	○	○	○
26.	○	○	○	○
27.	○	○	○	○
28.	○	○	○	○
29.	○	○	○	○
30.	○	○	○	○
31.	○	○	○	○
32.	○	○	○	○
33.	○	○	○	○

	A	B	C	D
34.	○	○	○	○
35.	○	○	○	○
36.	○	○	○	○
37.	○	○	○	○
38.	○	○	○	○
39.	○	○	○	○
40.	○	○	○	○
41.	○	○	○	○
42.	○	○	○	○
43.	○	○	○	○
44.	○	○	○	○
45.	○	○	○	○
46.	○	○	○	○
47.	○	○	○	○
48.	○	○	○	○
49.	○	○	○	○
50.	○	○	○	○
51.	○	○	○	○
52.	○	○	○	○
53.	○	○	○	○
54.	○	○	○	○
55.	○	○	○	○
56.	○	○	○	○
57.	○	○	○	○
58.	○	○	○	○
59.	○	○	○	○
60.	○	○	○	○
61.	○	○	○	○
62.	○	○	○	○
63.	○	○	○	○
64.	○	○	○	○
65.	○	○	○	○
66.	○	○	○	○
67.	○	○	○	○

	A	B	C	D
68.	○	○	○	○
69.	○	○	○	○
70.	○	○	○	○
71.	○	○	○	○
72.	○	○	○	○
73.	○	○	○	○
74.	○	○	○	○
75.	○	○	○	○
76.	○	○	○	○
77.	○	○	○	○
78.	○	○	○	○
79.	○	○	○	○
80.	○	○	○	○
81.	○	○	○	○
82.	○	○	○	○
83.	○	○	○	○
84.	○	○	○	○
85.	○	○	○	○
86.	○	○	○	○
87.	○	○	○	○
88.	○	○	○	○
89.	○	○	○	○
90.	○	○	○	○
91.	○	○	○	○
92.	○	○	○	○
93.	○	○	○	○
94.	○	○	○	○
95.	○	○	○	○
96.	○	○	○	○
97.	○	○	○	○
98.	○	○	○	○
99.	○	○	○	○
100.	○	○	○	○

Date _____

Score _____

Name _____

Chapter 7 Competency Profile

Student Name _____ Soc. Sec. No. _____

                Last   First   Middle

Fire Department_____

Address _____

Phone _____

Home Address _____

Phone _____

Date of Enrollment ____ - ____ - ____   Total Class Hours _____

Date of Withdrawal ____ - ____ - ____   Total Hours Absent_____

Date of Completion ____ - ____ - ____

Instructor's Name _____   Session Dates_____

Instructor's Directions

1. Check the candidate's competency rating (3, 2, 1, ☒) for each performance test task and psychomotor lesson objective (practical activity and job sheets) listed below.

2. List any additional performance tasks or psychomotor objectives (job sheets or practical activity sheets) under "Other," and check the candidate's competency rating.

3. Record the candidate's cognitive scores (written lesson tests and *administered* chapter review tests) in the spaces provided.

Level				Psychomotor Competencies
3	2	1	☒	

Lesson 7A — Rescue & Extrication Tools

Practical Activity Sheet

3	2	1	☒	
☐	☐	☐	☐	PAS 7A-1 — Select Correct Tools for Specific Situations
☐	☐	☐	☐	Other _____
☐	☐	☐	☐	_____

Job Sheets

3	2	1	☒	
☐	☐	☐	☐	JS 7A-1 — Safely Set Up Fire Service Lighting Equipment
☐	☐	☐	☐	JS 7A-2 — Service and Maintain Portable Power Plants and Lighting Equipment
☐	☐	☐	☐	JS 7A-3 — Use a Hydraulic Jack
☐	☐	☐	☐	JS 7A-4 — Use a Hydraulic Spreader
☐	☐	☐	☐	JS 7A-5 — Use Hydraulic Shears
☐	☐	☐	☐	JS 7A-6 — Use a Hydraulic Extension Ram

☐	☐	☐	☐	JS 7A-7 — Use a Bar Screw Jack
☐	☐	☐	☐	JS 7A-8 — Use a Trench Screw Jack
☐	☐	☐	☐	JS 7A-9 — Use a Ratchet-Lever Jack
☐	☐	☐	☐	JS 7A-10 — Use a Pneumatic Chisel/Hammer
☐	☐	☐	☐	JS 7A-11 — Use a Truck-Mounted Winch
☐	☐	☐	☐	JS 7A-12 — Use a Come-Along
☐	☐	☐	☐	JS 7A-13 — Use Air Lifting Bag(s)
☐	☐	☐	☐	JS 7A-14 — Use a Block and Tackle
☐	☐	☐	☐	JS 7A-15 — Use an Electric or Gasoline-Powered Circular Saw
☐	☐	☐	☐	JS 7A-16 — Use an Electric Reciprocating Saw
☐	☐	☐	☐	JS 7A-17 — Use an Electric or Gasoline-Powered Chain Saw
☐	☐	☐	☐	Other _____
☐	☐	☐	☐	_____

Lesson 7B — Vehicle Extrication & Special Rescue

Practical Activity Sheet

☐	☐	☐	☐	PAS 7B-1 — Assist Rescue Teams
☐	☐	☐	☐	Other _____
☐	☐	☐	☐	_____

Job Sheets

☐	☐	☐	☐	JS 7B-1 — Remove Automotive Window Glass
☐	☐	☐	☐	JS 7B-2 — Remove Vehicle Doors
☐	☐	☐	☐	JS 7B-3 — Move or Remove Vehicle Roofs
☐	☐	☐	☐	JS 7B-4 — Remove Steering Wheels and Columns
☐	☐	☐	☐	JS 7B-5 — Displace Dashboards
☐	☐	☐	☐	Other _____
☐	☐	☐	☐	_____

Chapter 7 Performance Test

☐	☐	☐	☐	Task 1 — Set up and operate power plants and lighting equipment to support operations.
☐	☐	☐	☐	Task 2 — Operate rescue and extrication equipment in support of operations.
☐	☐	☐	☐	Task 3 — Working as a team, extricate a victim from a vehicle.
☐	☐	☐	☐	Task 4 — Role-play rescue operations for special situations.

Psychomotor Competencies

Other _____

_____

Points Achieved	Points Needed/ Total	

Cognitive Competencies

Lesson 7A Written Test

Points Achieved	Points Needed/Total	
_____	6/8	1. Match facts about power plants to the equipment to which they apply.
_____	4/4	2. List the two types of lighting commonly used to support emergency operations.
_____	12/13	3. Complete statements regarding the care and use of auxiliary electrical equipment.
_____	4/4	4. Describe guidelines for maintaining power plants and lighting equipment.
		5. Evaluated on Job Sheet 7A-1
		6. Evaluated on Job Sheet 7A-2
_____	14/17	7. Identify rescue and extrication tools and equipment.
_____	4/5	8. Match hydraulic extrication and rescue tools to their purposes.
_____	6/6	9. List hydraulic tool safety guidelines.
		10. Evaluated on Job Sheets 7A-3 through 7A-6
_____	4/5	11. Match manual jacks and cribbing to their purposes.
_____	6/6	12. List jacking and cribbing safety guidelines.
		13. Evaluated on Job Sheets 7A-7 through 7A-9
_____	4/5	14. Match pneumatic rescue and extrication tools to their purposes.
_____	4/4	15. List pneumatic tool safety guidelines.
		16. Evaluated on Job Sheet 7A-10
_____	4/4	17. List winch safety guidelines.
		18. Evaluated on Job Sheet 7A-11
		19. Evaluated on Job Sheet 7A-12
_____	12/12	20. Complete air bag safety guidelines.
		21. Evaluated on Job Sheet 7A-13

Points Achieved	Points Needed/ Total	Cognitive Competencies
_____	7/8	22. Label the parts of a block and tackle.
_____	5/5	23. List block and tackle safety guidelines.
		24. Evaluated on Job Sheet 7A-14
		25. Evaluated on Job Sheets 7A-15 through 7A-17
		26. Evaluated on Practical Activity Sheet 7A-1

Lesson 7B Written Test

Points Achieved	Points Needed/ Total	
_____	5/5	1. List considerations to be made when sizing up a vehicle accident.
_____	6/8	2. List concerns of rescuers who assess the situation at automobile accidents.
_____	8/8	3. State the purpose of vehicle stabilization.
_____	6/6	4. List methods of gaining access to victims in vehicles.
_____	2/3	5. List complications of extrication efforts as a result of passenger restraint and protection systems.
_____	4/5	6. Select facts about disentanglement and patient management.
_____	8/8	7. State the purpose of packaging.
_____	5/6	8. Distinguish between *laminated glass* and *tempered glass*.
_____	2/3	9. Select the correct method for removing vehicle glass.
		10. Evaluated on Job Sheet 7B-1
_____	3/3	11. Match vehicle roof posts to their letter designations.
		12. Evaluated on Job Sheet 7B-2
		13. Evaluated on Job Sheet 7B-3
		14. Evaluated on Job Sheet 7B-4
		15. Evaluated on Job Sheet 7B-5
_____	6/8	16. Match types of building collapse to their descriptions.
_____	2/2	17. List the two types of hazards associated with structural collapse rescue operations.
_____	5/5	18. Distinguish between *shoring* and *tunneling*.
_____	5/6	19. Select facts about trench rescue operations.
_____	4/4	20. State the role of fire departments in cave and tunnel rescue operations.

Points Achieved	Points Needed/ Total	Cognitive Competencies
_____	5/5	21. Select facts about rescue operations involving electricity.
_____	5/6	22. Distinguish between *rescues* and *recoveries*.
_____	8/8	23. Describe the methods for performing a water rescue.
_____	6/6	24. Describe the methods for performing an ice rescue.
_____	4/4	25. List factors that should be taken into account during industrial extrications.
_____	4/5	26. Select facts about elevator and escalator rescues.
		27. Evaluated on Practical Activity Sheet 7B-1

Review Test

_____   Chapter 7 Review Test

Instructor's Signature _____ **Date** _____

Student's Signature _____ **Date** _____

STUDENT APPLICATIONS

FOURTH EDITION
ESSENTIALS OF FIRE FIGHTING

LESSON
11

HYDRANT FLOW & OPERABILITY

FIREFIGHTER II

FIRE PROTECTION PUBLICATIONS
OKLAHOMA STATE UNIVERSITY

Study Objectives

LESSON OBJECTIVE

After completing this lesson, you will be able to test the operability of and flow from a fire hydrant.

ENABLING OBJECTIVES

After reading Chapter 11 of **Essentials,** pages 385–389, and completing related activities, you will be able to —

1. Match to their correct definitions terms associated with water flow and pressure.

2. Select from a list conditions that reduce hydrant effectiveness.

3. **Measure and record hydrant flow pressures. *(Job Sheet 11-1)***

Study Sheet

Introduction This study sheet is intended to help you learn the Firefighter II material in Chapter 11 of *Essentials of Fire Fighting*, Fourth Edition. You may use it for self-study, or you may use it to review material that will be covered in the lesson and chapter review tests. The numbers in parentheses are the pages in *Essentials* on which the answers or terms can be found.

Chapter Vocabulary Be sure that you know the chapter-related meanings of the following terms and abbreviations. Use a dictionary or the glossary in *Fire Service Orientation and Terminology* if you cannot determine the meaning of the term from its context.

- Flow pressure *(386)*
- Normal operating pressure *(386)*
- Pitot tube *(389)*
- Residual pressure *(386)*
- Static pressure *(385, 386)*

Study Questions & Activities

1. List reasons for variation in hydrant flow. *(387)*

   _____

   _____

   _____

   _____

   _____

   _____

   _____

   _____

2. Complete the following table of NFPA hydrant color codes. *(Table 11.1, 388)*

Hydrant Class	Color	Flow
Class AA	a._____	1,500 gpm *(5 680 L/min)* or greater
b. _____	Green	c. _____ _____
Class B	d. _____	500–999 gpm *(1 900 L/min to 3 780 L/min)*
Class C	e. _____	f. _____ _____

3. Briefly define the four kinds of pressure with which the fire service is concerned. *(385, 386)*

   a. Static pressure _____

   _____

   _____

   b. Normal operating pressure _____

   _____

   _____

   c. Residual pressure _____

   _____

   _____

   d. Flow pressure _____

   _____

   _____

4. List problems that firefighters should look for when checking fire hydrants. *(388, 389)*

   _____   _____

   _____   _____

   _____   _____

   _____   _____

5. What is a pitot tube used for? *(389)* _____

6. How do handheld and fixed-mount pitot tubes differ? *(389)* _____

   _____

   _____

Information Sheet 11-1
Discharge Table for Circular Outlets

Outlet Pressure in psi	OUTLET DIAMETER IN INCHES											
	2⅜	2½	2⅝	2¾	2⅞	3	3⅛	3⅞	4	4⅜	4½	4⅝
	U.S. Gallons per Minute											
20	680	750	830	910	990	1080	1170	1800	1920	2290	2430	2570
22	710	790	870	950	1040	1130	1230	1890	2020	2400	2550	2700
24	740	820	910	1000	1090	1180	1290	1970	2110	2510	2660	2810
26	770	860	940	1040	1130	1230	1340	2050	2190	2620	2770	2930
28	800	890	980	1070	1170	1280	1390	2130	2280	2720	2880	3040
30	830	920	1010	1110	1210	1320	1430	2210	2350	2820	2980	3150
32	860	950	1050	1150	1260	1370	1480	2280	2430	2910	3080	3250
34	880	980	1080	1180	1290	1410	1530	2350	2510	3000	3170	3350
36	910	1010	1110	1220	1330	1450	1580	2420	2580	3080	3260	3440
38	930	1040	1140	1250	1370	1490	1620	2480	2650	3170	3350	3540
40	960	1060	1170	1290	1400	1530	1660	2550	2720	3250	3440	3630

*Computed with Coefficient C = 0.90, to nearest 10 gallons per minute.

DISCHARGE TABLE FOR CIRCULAR OUTLETS* (Metric)
Outlet Pressure Measured by Pitot Gauge

Outlet Pressure in kPa	OUTLET DIAMETER IN MILLIMETERS											
	60	64	67	70	73	76	79	98	102	111	114	117
	Liters per Minute											
20	968	1101	1206	1317	1432	1552	1677	2581	2796	3312	3493	3679
25	1082	1231	1348	1472	1601	1736	1875	2886	3126	3702	3905	4113
30	1185	1348	1478	1613	1754	1901	2054	3161	3425	4056	4278	4506
35	1280	1456	1596	1742	1894	2054	2219	3415	3699	4381	4620	4867
40	1368	1557	1706	1862	2026	2195	2372	3650	3954	4683	4940	5203
45	1451	1651	1810	1975	2148	2328	2516	3871	4194	4967	5239	5519
50	1530	1741	1907	2082	2264	2455	2652	4081	4421	5236	5523	5817
55	1604	1826	2001	2184	2375	2574	2781	4281	4637	5492	5792	6101
60	1676	1907	2090	2281	2481	2689	2905	4471	4843	5736	6050	6373
65	1744	1985	2175	2374	2582	2799	3024	4653	5041	5970	6293	6633
70	1810	2060	2257	2463	2679	2904	3138	4829	5231	6195	6535	6883
75	1873	2132	2336	2550	2773	3006	3248	4999	5415	6413	6764	7125
80	1935	2202	2413	2634	2864	3105	3355	5162	5593	6623	6986	7358
85	1994	2270	2487	2715	2952	3200	3458	5321	5765	6827	7201	7589
90	2053	2335	2559	2794	3038	3293	3558	5476	5932	7025	7410	7805
95	2109	2399	2629	2870	3121	3383	3656	5625	6094	7217	7612	8019
100	2164	2462	2698	2945	3203	3471	3751	5772	6253	7405	7810	8227
105	2217	2522	2764	3017	3282	3557	3843	5914	6407	7589	8003	8430
110	2269	2582	2860	3089	3359	3640	3933	6054	6558	7766	8192	8628
115	2320	2640	2893	3158	3434	3722	4022	6190	6705	7940	8376	8822
120	2370	2697	2955	3225	3508	3803	4109	6323	6849	8112	8556	9012
125	2419	2752	3016	3292	3581	3881	4193	6453	6990	8279	8732	9198
130	2467	2807	3076	3358	3652	3958	4277	6581	7129	8443	8905	9380
135	2514	2860	3135	3422	3721	4033	4358	6706	7265	8604	9075	9559
140	2560	2913	3192	3484	3789	4107	4438	6829	7398	8761	9241	9734
145	2605	2964	3249	3546	3856	4180	4516	6950	7529	8917	9405	9907
150	2650	3015	3304	3607	3922	4251	4594	7069	7658	9069	9569	10076

*Computed with Coefficient C = 0.90, to nearest liter.

Job Sheet 11-1
Measure and Record Hydrant Flow Pressures

Name _____ Date _____

Evaluator _____ Overall Competency Rating _____

References | NFPA 1001, Fireground Operations 4-5.4b
Essentials, pages 389, 394

Prerequisites | None

Student's Instructions | To meet evaluation standards, you must perform this job within _____ *[amount of time, if applicable]*; you may have _____ attempts. When you are ready to perform this job, ask your instructor to observe the procedure and complete this form. To show mastery of this job, you must perform all steps to receive an overall competency rating of at least 2.

Competency Rating Scale

3 — **Skilled** — Meets all evaluation criteria and standards; performs task independently on first attempt; requires no additional practice or training.

2 — **Moderately skilled** — Meets all evaluation criteria and standards; performs task independently; additional practice is recommended.

1 — **Unskilled** — Is unable to perform the task; additional training required.

☒ — **Unassigned** — Job sheet task is not required or has not been performed.

✔ **Evaluator's Note:** Formulate and inform the candidate of the standards for this task (time allowed and number of attempts). Observe the candidate perform the task, check the step/key point under the appropriate attempt number as accomplished, record total time (if appropriate), and then use the rating scale above to assign an overall competency rating. If the candidate is unable to perform any step of this job, have the candidate review the materials and try again.

Introduction | The rate of flow of water coming from a discharge opening produces a force that is called *flow pressure* or *velocity*. Because the stream of water as it emerges from the discharge opening is not encased in a tube, it does not exert an outward pressure. The forward velocity, or flow pressure, can be measured by using a pitot tube and gauge.

Flow pressures are measured in pounds per square inch *(psi)* or kilopascal *(kPa)*. The gallon per minute *(gpm)* or liters per minute *(L/min)* flow from the hydrant can be determined when the outlet diameter and flow pressure are known. The easiest way to determine how much water is flowing from the discharge opening is to refer to prepared tables. (See Information Sheet 11-1.) The gpm *(L/min)* flow can also be calculated, using various formulas.

✔**Note:** Remember to follow the hydrant safety rules learned in Firefighter I.

- Tighten caps on outlets not used.
- Do not stand in front of closed caps.
- Do not lean over top of operating hydrant.
- Close hydrant slowly.
- Check downstream drainage.
- Do not flow without adequate drainage.
- Do not flow across a busy street.
- Do not flow onto street in freezing weather.
- Control pedestrian and vehicle traffic.

Equipment and Personnel
- Two firefighters (one to take the readings and the other to record them)
- Handheld pitot tube and gauge
- Discharge Table for Circular Outlets *(Information Sheet 11-1)*
- Three different sized discharge outlets
- Spanner or hydrant wrench
- Hydrant logbook and pen or pencil for recording flow data

Job Steps	Key Points	Attempt No. 1	2	3
1. Take necessary safety precautions at the site.	1. a. Controlling pedestrian and vehicle traffic as necessary	___	___	___
	b. Checking downstream to see where water will flow	___	___	___
2. As a safety precaution, tighten those hydrant outlet caps that will not be used.	2. a. Turning caps clockwise	___	___	___
	b. Using spanner or hydrant wrench	___	___	___
3. Remove the cap from the outlet to be tested.	3. a. Turning outlet nut counterclockwise	___	___	___
	b. Standing clear of closed caps	___	___	___
4. Open the hydrant.	4. a. Slowly turning hydrant nut counterclockwise	___	___	___
	b. Using spanner or hydrant wrench	___	___	___
	c. Continuing until fully open	___	___	___
	d. Standing clear of closed caps	___	___	___
	e. Not leaning over top of hydrant	___	___	___
5. Open petcock at end of pitot tube air chamber.	5. a. Draining all water from assembly	___	___	___
	b. Closing petcock	___	___	___

Job Steps	Key Points	1	2	3
6. Position yourself to take reading.	6. a. To right or left side of stream	—	—	—
	b. Not leaning over top of hydrant	—	—	—
7. Grasp pitot tube.	7. a. Fingers and thumb of one hand just behind blade	—	—	—
	b. Other hand firmly gripping air chamber (handle) **or**	—	—	—
	a. Splitting fingers of one hand around gauge outlet (palm down, gauge face up)	—	—	—
	b. Other hand firmly gripping air chamber (handle)	—	—	—
8. Position the pitot tube in the flow stream.	8. a. At a right angle to the stream	—	—	—
	b. Gauge facing up	—	—	—
	c. Pitot point (small opening) in center of stream	—	—	—
	d. Holding at a distance away from threads equal to about half the diameter of outlet opening	—	—	—
	✔**Note:** For a 2½-inch *(65 mm)* hydrant butt, this distance would be approximately 1½ inches *(32 mm).*			
	e. Keeping air chamber above horizontal plane through center of stream	—	—	—
	f. Steadying tube by placing little finger of "lade" hand on top of hydrant outlet or by resting side of fist against hydrant outlet	—	—	—
9. Read pitot tube gauge.	9. Taking reading between high and low values if needle is fluctuating	—	—	—
10. Report the flow pressure reading.	10. To second firefighter with logbook	—	—	—

			Attempt No.		
Job Steps	Key Points		1	2	3
11. *(Second firefighter)* Record the flow pressure reading.	11. In hydrant logbook		—	—	—
12. Close the hydrant.	12. a. Slowly turning hydrant nut clockwise		—	—	—
	b. Using spanner or hydrant wrench		—	—	—
	c. Continuing until fully closed		—	—	—
	d. Standing clear of closed caps		—	—	—
13. Replace cap on tested outlet.	13. a. Turning outlet nut clockwise until firmly closed		—	—	—
	b. Standing clear of closed caps		—	—	—
14. Open petcock at end of pitot tube air chamber.	14. a. Draining all water from assembly		—	—	—
	b. Closing petcock		—	—	—
15. Repeat the procedure.	15. On outlets of two different sizes		—	—	—
16. Convert psi readings to gpm or kPa to L/min.	16. Using Information Sheet 11-1 tables		—	—	—
17. Record converted readings in logbook.	17. (None)		—	—	—

Time (Total)   — — —

Evaluator's Comments _____

_____

_____

Chapter 11 Review Test

> → **Directions:** This review test covers the Firefighter II material in Chapter 11 of your ***Essentials of Fire Fighting*** text. It may be assigned as a study aid (self-test) or may be administered by your instructor as a pretest or posttest.
>
> When used as a study aid, try to answer the questions without referring to the page numbers in ***Essentials*** or your ***Firefighter II Student Applications*** workbook *(SA)* on which the answers can be found until after you have completed the entire test. Then check your answers against those on the pages provided in parentheses.
>
> When administered by your instructor as a pretest or posttest, read each of the test questions carefully. Choose the best response and then darken the corresponding letter on your answer sheet.
>
> This chapter review test contains 10 multiple-choice questions, each worth 10 points. To pass the test, you must achieve at least 80 of the 100 points possible.

1. What is the most common fire service definition of *pressure? (385)*

    A. Compression of water volume

    B. Force of water per unit of area

    C. Quickness of water motion

    D. Velocity of water in a conduit

2. How is pressure — in the fire service sense — measured? *(385)*

    A. Pounds per square inch or kilopascals

    B. Miles or kilometers per hour

    C. Gallons or liters per minute

    D. Gallons per square foot or liters per square meter

3. What is the best definition of *static pressure? (386)*

    A. That part of the total available pressure that is not used to overcome friction or gravity while forcing water through pipe, fittings, fire hose, and adapters

    B. Pressure found in a water distribution system during periods of normal consumption demand

    C. Forward velocity pressure at a water distribution system discharge opening (either hydrant or nozzle) while water is flowing

    D. Stored potential energy that is available to force water through pipe, fittings, fire hose, and adapters

4. How is *residual pressure* defined? *(386)*

    A. Stored potential energy that is available to force water through pipe, fittings, fire hose, and adapters

    B. Pressure found in a water distribution system during periods of normal consumption demand

    C. Forward velocity pressure at a water distribution system discharge opening (either hydrant or nozzle) while water is flowing

    D. That part of the total available pressure that is not used to overcome friction or gravity while forcing water through pipe, fittings, fire hose, and adapters

5. What is the best definition of *normal operating pressure?* *(386)*

   A. Stored potential energy that is available to force water through pipe, fittings, fire hose, and adapters

   B. That part of the total available pressure that is not used to overcome friction or gravity while forcing water through pipe, fittings, fire hose, and adapters

   C. Pressure found in a water distribution system during periods of normal consumption demand

   D. Forward velocity pressure at a water distribution system discharge opening (either hydrant or nozzle) while water is flowing

6. Which of the following best describes *flow pressure?* *(386)*

   A. Stored potential energy that is available to force water through pipe, fittings, fire hose, and adapters

   B. That part of the total available pressure that is not used to overcome friction or gravity while forcing water through pipe, fittings, fire hose, and adapters

   C. Forward velocity pressure at a water distribution system discharge opening (either hydrant or nozzle) while water is flowing

   D. Pressure found in a water distribution system during periods of normal consumption demand

7. Which of the following does **not** directly affect hydrant water flow? *(387)*

   A. Softness of water

   B. Sedimentation and mineral deposits within the distribution system

   C. Size of main to which hydrant is connected

   D. Proximity of feeder mains

8. In general, hydrants in high-value districts should not be placed more than ___ apart. *(388)*

   A. 200 feet *(60 m)*   C. 500 feet *(150 m)*

   B. 300 feet *(90 m)*   D. One block

9. Class C fire hydrants are painted ___ and have a flow of ___. *(Table 11.1, 388)*

   A. Red; less than 500 gpm *(1 900 L/min)*

   B. Green; 1,000 to 1,499 gpm *(3 785 L/min to 5 675 L/min)*

   C. Light blue; 1,500 gpm *(5 680 L/min)*

   D. Orange; 500 to 999 gpm *(1 900 L/min to 3 780 L/min)*

10. According to the NFPA hydrant color code, which class of fire hydrant has the least powerful water flow? *(Table 11.1, 318)*

    A Class AA   C. Class B

    B. Class A   D. Class C

REVIEW TEST ANSWER SHEET

	A	B	C	D
1.	○	○	○	○
2.	○	○	○	○
3.	○	○	○	○
4.	○	○	○	○
5.	○	○	○	○
6.	○	○	○	○
7.	○	○	○	○
8.	○	○	○	○
9.	○	○	○	○
10.	○	○	○	○
11.	○	○	○	○
12.	○	○	○	○
13.	○	○	○	○
14.	○	○	○	○
15.	○	○	○	○
16.	○	○	○	○
17.	○	○	○	○
18.	○	○	○	○
19.	○	○	○	○
20.	○	○	○	○
21.	○	○	○	○
22.	○	○	○	○
23.	○	○	○	○
24.	○	○	○	○
25.	○	○	○	○
26.	○	○	○	○
27.	○	○	○	○
28.	○	○	○	○
29.	○	○	○	○
30.	○	○	○	○
31.	○	○	○	○
32.	○	○	○	○
33.	○	○	○	○

	A	B	C	D
34.	○	○	○	○
35.	○	○	○	○
36.	○	○	○	○
37.	○	○	○	○
38.	○	○	○	○
39.	○	○	○	○
40.	○	○	○	○
41.	○	○	○	○
42.	○	○	○	○
43.	○	○	○	○
44.	○	○	○	○
45.	○	○	○	○
46.	○	○	○	○
47.	○	○	○	○
48.	○	○	○	○
49.	○	○	○	○
50.	○	○	○	○
51.	○	○	○	○
52.	○	○	○	○
53.	○	○	○	○
54.	○	○	○	○
55.	○	○	○	○
56.	○	○	○	○
57.	○	○	○	○
58.	○	○	○	○
59.	○	○	○	○
60.	○	○	○	○
61.	○	○	○	○
62.	○	○	○	○
63.	○	○	○	○
64.	○	○	○	○
65.	○	○	○	○
66.	○	○	○	○
67.	○	○	○	○

	A	B	C	D
68.	○	○	○	○
69.	○	○	○	○
70.	○	○	○	○
71.	○	○	○	○
72.	○	○	○	○
73.	○	○	○	○
74.	○	○	○	○
75.	○	○	○	○
76.	○	○	○	○
77.	○	○	○	○
78.	○	○	○	○
79.	○	○	○	○
80.	○	○	○	○
81.	○	○	○	○
82.	○	○	○	○
83.	○	○	○	○
84.	○	○	○	○
85.	○	○	○	○
86.	○	○	○	○
87.	○	○	○	○
88.	○	○	○	○
89.	○	○	○	○
90.	○	○	○	○
91.	○	○	○	○
92.	○	○	○	○
93.	○	○	○	○
94.	○	○	○	○
95.	○	○	○	○
96.	○	○	○	○
97.	○	○	○	○
98.	○	○	○	○
99.	○	○	○	○
100.	○	○	○	○

Name _____

Date _____
Score _____

Chapter 11 Competency Profile

Student Name _____ Soc. Sec. No. _____
　　　　　　　Last　　　　　First　　　　　Middle

Fire Department_____

Address _____

Phone _____

Home Address _____

Phone _____

Date of Enrollment ____ - ____ - ____　　　　Total Class Hours _____

Date of Withdrawal ____ - ____ - ____　　　　Total Hours Absent_____

Date of Completion ____ - ____ - ____

Instructor's Name _____　　　Session Dates_____

Instructor's Directions

1. Check the candidate's competency rating (3, 2, 1, ☒) for each performance test task and psychomotor lesson objective (practical activity and job sheets) listed below.

2. List any additional performance tasks or psychomotor objectives (job sheets or practical activity sheets) under "Other," and check the candidate's competency rating.

3. Record the candidate's cognitive scores (written lesson tests and *administered* chapter review tests) in the spaces provided.

Level	Psychomotor Competencies

3	2	1	☒

Practical Activity Sheets

None Required

☐	☐	☐	☐	Other _____
☐	☐	☐	☐	_____

Job Sheets

☐	☐	☐	☐	JS 11-1 — Measure and Record Hydrant Flow Pressures
☐	☐	☐	☐	Other _____
☐	☐	☐	☐	_____

Level
3 2 1 ☒

☐	☐	☐	☐
☐	☐	☐	☐
☐	☐	☐	☐

Psychomotor Competencies

Chapter 11 Performance Test

Task 1 — Test the operability and flow from a fire hydrant.

Other _____

_____

Points Achieved	Points Needed/ Total

Cognitive Competencies

Written Test

_____ 30/40	1. Match to their correct definitions terms associated with water flow and pressure.
_____ 48/60	2. Select from a list conditions that reduce hydrant effectiveness.
_____	3. Evaluated on Job Sheet 11-1

Review Test

_____ Chapter 11 Review Test

Instructor's Signature _____ **Date** _____

Student's Signature _____ **Date** _____

STUDENT APPLICATIONS

FOURTH EDITION

ESSENTIALS OF FIRE FIGHTING

LESSON

12

HOSE TOOLS & APPLIANCES

FIREFIGHTER II

FIRE PROTECTION PUBLICATIONS
OKLAHOMA STATE UNIVERSITY

Study Objectives

LESSON OBJECTIVE

After completing this lesson, you will be able to identify and use hose tools and appliances and service test hose.

ENABLING OBJECTIVES

After reading **Essentials,** Chapter 12, pages 406–413, 437, and 438 and completing related activities, you will be able to —

1. Identify types of valves and valve devices.

2. Match types of valves to their functions.

3. Identify hose fitting appliances.

4. Identify tools used with hose.

5. Match hose appliances and tools to their uses in specific fireground situations.

6. **Select adapters and appliances for given fireground situations. (Practical Activity Sheet 12-1)**

7. **Use hose tools and appliances. (Job Sheets 12-1 – 12-6)**

8. Select facts about service testing hose.

9. List safety guidelines for service testing hose.

10. **Service test hose. (Job Sheet 12-7)**

Study Sheet

Introduction

This study sheet is intended to help you learn the material in Chapter 12 of ***Essentials of Fire Fighting***, Fourth Edition, pages 406–413, 437, and 438. You may use it for self-study, or you may use it to review material that will be covered in the lesson and chapter review tests. The numbers in parentheses are the pages in ***Essentials*** on which the answers or terms can be found.

Chapter Vocabulary

Be sure that you know the chapter-related meanings of the following terms and abbreviations. Use a dictionary or the glossary in ***Fire Service Orientation and Terminology*** if you cannot determine the meaning of the term from its context:

- Baffle *(406)*
- Male *(407)*
- Female *(407)*

- Chafing *(413)*
- Hose appliances *vs.* hose tools *(406)*
- Hose caps *vs.* hose plugs *(410)*

Study Questions & Activities

1. Identify the following hose tools and appliances. Write the correct name of the appliance below its illustration, and then briefly state the function of the appliance or tool.

a. _____
   _____ *(406, 407)*

b. _____
   _____ *(406, 407)*

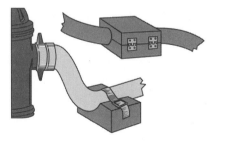

c. _____
   _____ *(413)*

d. _____
   _____ *(406, 407)*

e. _____
_____ *(409)*

f. _____
_____ *(409)*

g. _____
_____ *(407)*

h. _____
_____ *(411)*

i. _____
_____ *(411)*

j. _____
_____ *(412)*

k. _____
_____ *(410)*

l. _____
_____ *(413)*

m. _____
_____ *(407)*

n. _____
_____ *(404)*

o. _____   p. _____

_____ (408)   _____ (407)

2. What items are included in a complete hose layout? In the hose appliances category? In the hose tools category? *(406)*

   a. Hose layout _____

   _____

   b. Hose appliances _____

   _____

   c. Hose tools _____

   _____

3. Describe the functions of the following types of valves, and identify where they are used. *(406)*

   a. Ball valves _____

   Where used _____

   b. Gate valves _____

   Where used _____

   c. Butterfly valves _____

   Where used _____

   d. Clapper valves _____

   Where used _____

4. Briefly describe the following valve devices. *(406–409)*

   a. Wye appliances _____

   _____

   _____

   b. Siamese appliances _____

   _____

   _____

   c. Water thief appliances _____

   _____

   _____

   d. Large diameter hose appliances _____

   _____

   _____

e. Hydrant valves _____

_____

_____

5. What is the purpose of fittings? Describe two examples. *(409, 410)*

a. Purpose _____

_____

b. Examples _____

_____

_____

_____

6. What is a suction hose strainer? Name two types of suction hose strainers. *(410)*

a. Definition _____

_____

_____

b. Types _____

7. What hose tool is used to prevent damage to a hose when it must be dragged over sharp surfaces? How is it used? *(410)*

_____

_____

8. What is a hose jacket, and when is it used? *(410, 411)*

_____

_____

_____

9. Name three situations in which hose clamps are used to stop the flow of water through a hoseline. *(411)*

a. _____

_____

b. _____

_____

c. _____

_____

10. Name three types of hose clamps. *(411)*

a. _____

b. _____

c. _____

11. List general rules for using hose clamps. *(411, 412)*

_____

_____

_____

_____

12. Name the primary purpose of a spanner wrench and list some of the other features built into it. *(412)*

a. Primary _____

_____

b. Other _____

_____

_____

13. List the primary purpose of the following hose tools. *(412, 413)*

a. Hydrant wrench _____

_____

b. Rubber mallet _____

_____

c. Hose bridge _____

_____

d. Chafing block _____

_____

e. Hose strap _____

_____

14. Name the two types of tests for fire hose. Which is done by the manufacturer before the hose is shipped? *(437)*

_____

_____

15. When should fire department hose be service tested? *(437)*

_____

16. List equipment needed to service test hose. *(437)*

_____

_____

_____

_____

_____

17. List safety precautions that should be taken when service testing hose. *(437, 438)*

_____

_____

_____

_____

18. List items that are checked prior to and during service testing of hose. *(437)*

_____

_____

_____

_____

Practical Activity Sheet 12-1
Select Adapters and Appliances for
Given Fireground Situations

Name _____ Date _____

Evaluator _____ Overall Competency Rating _____

Reference	NFPA 1001, Fireground Operations 4-3.2a
Prerequisites	*Essentials*, Chapter 11
Introduction	On the fireground, when the firefighter is faced with mismatched couplings, a burst hose, a coupling that leaks excessively and cannot be further tightened, hose that must be protected from vehicles, lines that need to be divided, or any number of other problems, it is essential that the firefighter know what adapters and appliances to use to connect, protect, replace, and repair these working hoselines.
Directions	Select the correct adapter and/or appliance for the situations described below. Write the correct answers on the blanks.
Activity	**SITUATION 1:** A section of hose has ruptured and you are unable to shut down the line and replace the damaged section. Adapter/appliance selected _____ **SITUATION 2:** You need to connect two sections of hose, but both couplings are male. Adapter/appliance selected _____ **SITUATION 3:** You need to extend a section of hoseline without shutting down the water supply. Adapter/appliance selected _____ **SITUATION 4:** You are making a forward lay, and need to connect the line to the hydrant and charge it before the arrival of another pumper at the hydrant. Adapter/appliance selected _____ **SITUATION 5:** You need to extend a 2½-inch *(65 mm)* line with a 1½-inch *(38 mm)* line. Adapter/appliance selected _____

Situation 6: You need to supply water to a ladder pipe that is not equipped with a waterway.

Adapter/appliance selected _____

Situation 7: You need to bring two or more hoselines into one hoseline.

Adapter/appliance selected _____

Situation 8: You need to divide a hoseline into two or more lines.

Adapter/appliance selected _____

Situation 9: You need to branch a number of 2½-inch *(65 mm)* lines off LDH.

Adapter/appliance selected _____

Situation 10: For mop-up after a grass fire, you want to branch a 1-inch *(25 mm)* line off a 1½-inch *(38 mm)* line.

Adapter/appliance selected _____

Practical Evaluation

PAS 12-1 —Select Adapters and Appliances for Given Fireground
             Situations

Competency Rating Scale
3 — Skilled — All 10 situations answered correctly. **2 — Moderately skilled** — At least 7 of the 10 situations answered correctly; additional practice is recommended. **1 — Unskilled** — Fewer than 7 of the 10 situations answered correctly. ☒ —**Unassigned** — Task is not required or has not been performed. ✔ **Evaluator's Note:** Evalute the results or product as indicated below. Calculate the percentage of correct responses, and assign an averall competency level. Record this competency rating on both the student's practical activity sheet and competency profile. To show competency in this objective, the student must achieve an overall rating of at least 2.

Criteria/Answers	Correct	Incorrect
*All answers evaluated per answers in **Instructor's Guide.***		
Situation 1	☐	☐
Situation 2	☐	☐
Situation 3	☐	☐
Situation 4	☐	☐
Situation 5	☐	☐
Situation 6	☐	☐
Situation 7	☐	☐
Situation 8	☐	☐
Situation 9	☐	☐
Situation 10	☐	☐

Job Sheet 12-1
Apply a Hose Jacket to a Leaking Coupling

Name _____ **Date** _____

Evaluator _____ **Overall Competency Rating** _____

References	NFPA 1001, Fireground Operations 4-3.1, 4-3.2, 4-3.3 *Essentials*, pages 410, 411
Prerequisites	None
Student's Instructions	To meet evaluation standards, you must perform this job within _____ *[amount of time, if applicable]*; you may have _____ attempts. When you are ready to perform this job, ask your instructor to observe the procedure and complete this form. To show mastery of this job, you must perform all steps to receive an overall competency rating of at least 2.

> **Competency Rating Scale**
>
> **3 — Skilled** — Meets all evaluation criteria and standards; performs task independently on first attempt; requires no additional practice or training.
>
> **2 — Moderately skilled** — Meets all evaluation criteria and standards; performs task independently; additional practice is recommended.
>
> **1 — Unskilled** — Is unable to perform the task; additional training required.
>
> ☒ **— Unassigned** — Job sheet task is not required or has not been performed.
>
> ✔ **Evaluator's Note:** Formulate and inform the candidate of the standards for this task (time allowed and number of attempts). Observe the candidate perform the task, check the step/key point under the appropriate attempt number as accomplished, record total time (if appropriate), and then use the rating scale above to assign an overall competency rating. If the candidate is unable to perform any step of this job, have the candidate review the materials and try again.

Introduction	The hose jacket is a tool used to seal small cuts or breaks that may occur in fire hose or to connect mismatched or damaged couplings of the same size.
Equipment and Personnel	• One firefighter in protective clothing • Hose jacket • Charged hoseline with leaking coupling

Job Steps	Key Points	Attempt No.		
		1	2	3
1. Open and position the hose jacket.	1. a. One side under hose and leaking couplings	___	___	___
	b. Edges of any flattened hose in jacket	___	___	___
	c. Upper half closed but not locked	___	___	___
2. Lock both sides of the hose jacket.	2. Stomping on it with one foot	___	___	___
3. Ensure that jacket is locked and seal is watertight.	3. a. Inspecting visually	___	___	___
	b. Removing and reapplying as necessary	___	___	___
	Time (Total)	___	___	___

Evaluator's Comments _____

_____

_____

Job Sheet 12-2
Replace a Section of Hose in a Charged Line

Name _____ Date _____

Evaluator _____ Overall Competency Rating _____

References	NFPA 1001, Fireground Operations, 4-3.1, 4-3.2, 4-3.3 *Essentials*, pages 411, 412
Prerequisites	Firefighter I hose competencies in coupling and uncoupling hose and unloading and carrying hose
Student's Instructions	To meet evaluation standards, you must perform this job within _____ *[amount of time, if applicable]*; you may have _____ attempts. When you are ready to perform this job, ask your instructor to observe the procedure and complete this form. To show mastery of this job, you must perform all steps to receive an overall competency rating of at least 2.

Competency Rating Scale

3 — Skilled — Meets all evaluation criteria and standards; performs task independently on first attempt; requires no additional practice or training.

2 — Moderately skilled — Meets all evaluation criteria and standards; performs task independently; additional practice is recommended.

1 — Unskilled — Is unable to perform the task; additional training required.

☒ — **Unassigned** — Job sheet task is not required or has not been performed.

✔ **Evaluator's Note:** Formulate and inform the candidate of the standards for this task (time allowed and number of attempts). Observe the candidate perform the task, check the step/key point under the appropriate attempt number as accomplished, record total time (if appropriate), and then use the rating scale above to assign an overall competency rating. If the candidate is unable to perform any step of this job, have the candidate review the materials and try again.

Introduction	There are three methods of shutting off water flow to a hoseline: at the apparatus, with a hose clamp, or by using the two-person kink. This job sheet outlines the procedures for using a hose clamp and for using the two-person kink. The methods of applying a hose clamp to a section of fire hose vary somewhat according to the kind of clamp used. The clamping jaws of most hose clamps are designed to prevent excessive pressure on the hose and subsequent injury to the hose jacket and lining. Some clamps are adjustable to fit several sizes of hose. The general types of hose clamps may be classified by the method in which they are operated: press-down lever clamp, screw-down clamp, and hydraulic clamp.

While a hose clamp is the best means of cutting off water flow, tools may not be available at the point of failure. Because the need for immediate shut-off is usually urgent, the quickest method is the two-person kink.

✔ **Note:** This job sheet should not be performed on an actual burst hoseline. Your instructor will have marked a chalk "X" on the line to indicate the site of the "break."

Equipment and Personnel

• Three firefighters in protective clothing, two to perform the kink and one to replace the hose section
• Charged 2½-inch *(65 mm)* hoseline
• Two additional sections of 2½-inch *(65 mm)* hose

✔ **Note:** Hoselines stretch to longer lengths when under pressure; thus, the couplings in the charged line will be farther apart than the length of a single replacement section.

• Hose clamp
• Spanner wrenches

Job Steps	Key Points	Attempt No. 1 2 3
HOSE CLAMP METHOD		
1. Determine the proper location for the hose clamp.	1. a. On supply side of break	__ __ __
	b. Not closer than 5 feet *(1.5 m)* from coupling	__ __ __
2. Place the hose in the clamp.	2. Centered in jaws to prevent pinching	__ __ __
3. Close the hose clamp jaws.	3. a. Standing to one side	__ __ __
	b. Slowly to prevent water hammer	__ __ __
4. Carry two new sections of hose.	4. a. From apparatus	__ __ __
	b. To coupling between clamp site and point of failure	__ __ __
	c. Using an approved unload and carry per departmental SOPs	__ __ __
5. Uncouple the damaged section of hose.	5. a. Using foot-tilt or over-hip method	__ __ __
	b. Turning counterclockwise	__ __ __
6. Connect a new section of hose.	6. a. To coupling on supply side of failure	__ __ __
	b. Turning clockwise	__ __ __
	c. Until hand-tight	__ __ __

Job Steps	Key Points	1	2	3
7. Connect the second new section of hose.	7. a. To first section	—	—	—
	b. Turning clockwise	—	—	—
	c. Until hand-tight	—	—	—
8. Connect opposite end of the second section.	8. a. To original line	—	—	—
	b. Turning clockwise	—	—	—
	c. Until hand-tight	—	—	—
9. Release the hose clamp.	9. a. Slowly to prevent violent movement of hose	—	—	—
	b. Watching for quick movement of hose as it is charged	—	—	—
10. Tighten connections as necessary.	10. Using spanner wrenches	—	—	—
11. Return clamp and wrenches to proper apparatus storage.	11. (None)	—	—	—

Time (Total)   —   —   —

Evaluator's Comments _____

_____

_____

TWO-PERSON KINK METHOD				
1. Determine the proper location for the kink.	1. a. On supply side of break	—	—	—
	b. Not closer than 5 feet (1.5 m) from coupling	—	—	—
2. Prepare the damaged hoseline.	2. Creating sufficient slack in line between point of failure and water supply source	—	—	—
3. Carry two new sections of hose.	3. a. From apparatus	—	—	—
	b. To coupling between kink site and point of failure	—	—	—
	c. Using an approved unload and carry per departmental SOPs	—	—	—

Job Steps	Key Points	Attempt No. 1 2 3

Job Steps	Key Points	1	2	3
4. Lay out the two new sections of hose.	4. a. Within reach	—	—	—
	b. Appropriate ends (male or female, depending on lay) toward water supply	—	—	—
5. Bend the hose.	5. a. Two firefighters	—	—	—
	b. Forming a round turn	—	—	—
	c. Loop perpendicular to ground	—	—	—
6. *(Two firefighters)* Apply pressure to hose loop.	6. a. Using body weight	—	—	—
	b. Fattening the loop	—	—	—
	c. One knee directly on each side of bend	—	—	—
	d. Shutting off water supply	—	—	—
	e. Maintaining pressure while new hose sections are added by third firefighter	—	—	—
(Third firefighter) 7. Uncouple the damaged section of hose.	7. a. When pressure has cut off water flow	—	—	—
	b. Using foot-tilt or over-hip method	—	—	—
	c. Turning counterclockwise	—	—	—
8. Connect a new section of hose.	8. a. To coupling on supply side of failure	—	—	—
	b. Turning clockwise	—	—	—
	c. Until hand-tight	—	—	—
9. Connect the second new section of hose.	9. a. To first section	—	—	—
	b. Turning clockwise	—	—	—
	c. Until hand-tight	—	—	—
10. Connect opposite end of the second section.	10. a. To original line	—	—	—
	b. Turning clockwise	—	—	—
	c. Until hand-tight	—	—	—
11. *(Two firefighters on kink)* Release the pressure on the hoseline.	11. a. Slowly to prevent violent movement of hose	—	—	—
	b. Watching for quick movement of hose as it is charged	—	—	—

Job Steps	Key Points	Attempt No.		
		1	2	3
(Third firefighter)				
12. Tighten connections as necessary.	12. Using spanner wrenches	—	—	—
13. Return spanner wrenches to proper storage on apparatus.	13. (None)	—	—	—
	Time (Total)	—	—	—

Evaluator's Comments _____

_____

_____

Job Sheet 12-3
Extend a Charged Hoseline

Name _____ Date _____

Evaluator _____ Overall Competency Rating _____

References	NFPA 1001, Fireground Operations 4-3.1, 4-3.2 *Essentials*, pages 411, 412, and 479
Prerequisites	Firefighter I hose competencies in coupling and uncoupling hose and unloading and carrying hose
Student's Instructions	To meet evaluation standards, you must perform this job within _____ *[amount of time, if applicable]*; you may have _____ attempts. When you are ready to perform this job, ask your instructor to observe the procedure and complete this form. To show mastery of this job, you must perform all steps to receive an overall competency rating of at least 2.

> **Competency Rating Scale**
>
> **3 — Skilled** — Meets all evaluation criteria and standards; performs task independently on first attempt; requires no additional practice or training.
>
> **2 — Moderately skilled** — Meets all evaluation criteria and standards; performs task independently; additional practice is recommended.
>
> **1 — Unskilled** — Is unable to perform the task; additional training required.
>
> ☒ **— Unassigned** — Job sheet task is not required or has not been performed.
>
> ✔ **Evaluator's Note:** Formulate and inform the candidate of the standards for this task (time allowed and number of attempts). Observe the candidate perform the task, check the step/key point under the appropriate attempt number as accomplished, record total time (if appropriate), and then use the rating scale above to assign an overall competency rating. If the candidate is unable to perform any step of this job, have the candidate review the materials and try again.

Introduction	During fire fighting operations, it may become necessary to extend a charged hoseline. In this job sheet, you learn how to extend the length of a charged hoseline by adding additional sections. It is also possible to extend a section of hose by using a break-apart nozzle.
Equipment and Personnel	• One firefighter in protective clothing • One charged 2½-inch *(65 mm)* line with nozzle • One section of 2½-inch *(65 mm)* hose • Hose clamp

Job Steps	Key Points	1	2	3
1. Close the nozzle on charged hose.	1. Slowly to prevent water hammer	—	—	—
2. Carry a section of hose and hose clamp from apparatus to nozzle end of charged line.	2. Using an approved unload and carry per departmental SOPs	—	—	—
3 Relieve nozzle pressure.	3. Opening slightly	—	—	—
4. Shut down the line by applying the hose clamp.	4. a. Approximately 3 feet *(1 m)* behind nozzle	—	—	—
	b. Hose in center of clamp	—	—	—
5. Remove the nozzle.	5. a. Using foot-tilt or over-hip method	—	—	—
	b. Laying it where it will not be a tripping hazard	—	—	—
6. Connect the new section of hose.	6. a. Female end to male end of clamped line	—	—	—
	b. Turning clockwise	—	—	—
	c. Until hand-tight	—	—	—
7. Reattach the nozzle.	7. a. To male coupling of extended hose	—	—	—
	b. Turning clockwise	—	—	—
	c. Until hand-tight	—	—	—
	d. Bale closed	—	—	—
8. Release the hose clamp.	8. a. From the original line	—	—	—
	b. Slowly to prevent violent movement of hose	—	—	—
9. Return clamp and wrenches to proper storage on apparatus.	9. (None)	—	—	—

	Time (Total)	—	—	—

Evaluator's Comments _____

_____

_____

Job Sheet 12-4
Reduce a Charged Hoseline

Name _____ Date _____

Evaluator _____ Overall Competency Rating _____

Reference	NFPA 1001, Fireground Operations 4-3.1, 4-3.2, 4-3.3
Prerequisites	Firefighter I hose competencies in coupling and uncoupling hose and unloading and carrying hose
Student's Instructions	To meet evaluation standards, you must perform this job within _____ *[amount of time, if applicable]*; you may have _____ attempts. When you are ready to perform this job, ask your instructor to observe the procedure and complete this form. To show mastery of this job, you must perform all steps to receive an overall competency rating of at least 2.

Competency Rating Scale

3 — Skilled — Meets all evaluation criteria and standards; performs task independently on first attempt; requires no additional practice or training.

2 — Moderately skilled — Meets all evaluation criteria and standards; performs task independently; additional practice is recommended.

1 — Unskilled — Is unable to perform the task; additional training required.

☒ **—Unassigned** — Job sheet task is not required or has not been performed.

✔ **Evaluator's Note:** Formulate and inform the candidate of the standards for this task (time allowed and number of attempts). Observe the candidate perform the task, check the step/key point under the appropriate attempt number as accomplished, record total time (if appropriate), and then use the rating scale above to assign an overall competency rating. If the candidate is unable to perform any step of this job, have the candidate review the materials and try again.

Introduction	Lines need to be reduced for a variety of reasons, most frequently because smaller lines are easier to handle. After fires are knocked down with larger lines, it is common practice to reduce to smaller lines for salvage and overhaul.
Equipment and Personnel	• One firefighter in protective clothing • One section of charged 2½-inch *(65 mm)* hose/nozzle • One section of 1½-inch *(38 mm)* hose/nozzle • One 2½- to 1½-inch *(65 mm to 38 mm)* reducer or gated wye • Hose clamp (optional)

Job Steps	Key Points	Attempt No.		
		1	2	3
1. Carry the section of 1½-inch *(38 mm)* hose and reducer or wye from apparatus to nozzle end of charged line.	1. Using an approved unload and carry per departmental SOPs	___	___	___
2. Shut down the line.	2. a. At source, *or*	___	___	___
	b. Hose clamp 3 to 5 feet *(1 m to 1.5 m)* behind nozzle	___	___	___
3. Open the nozzle to relieve pressure.	3. (None)	___	___	___
4. Remove the nozzle.	4. a. Using foot-tilt or over-hip method	___	___	___
	b. Laying nozzle where it will not be a tripping hazard	___	___	___
5. Attach the reducer or wye.	5. a. Using foot-tilt method	___	___	___
	b. Turning clockwise	___	___	___
	c. Until hand-tight	___	___	___
6. Attach the 1½-inch *(38 mm)* hose with nozzle.	6. a. Using foot-tilt method	___	___	___
	b. Nozzle closed	___	___	___
	c. Turning clockwise	___	___	___
	d. Until hand-tight	___	___	___
7. Charge the line.	7. a. At source or by removing clamp	___	___	___
	b. Slowly to prevent violent movement of hose	___	___	___
	Time (Total)	___	___	___

Evaluator's Comments _____

_____

_____

Job Sheet 12-5
Siamese Two or More Lines
Into One Line

Name _____ Date _____

Evaluator _____ Overall Competency Rating _____

References	NFPA 1001, Fireground Operations 4-3.1, 4-3.2, 4-3.2a, 4-3.3 *Essentials*, pages 407, 408
Prerequisites	Firefighter I hose competencies in coupling and uncoupling hose and unloading and carrying hose
Student's Instructions	To meet evaluation standards, you must perform this job within _____ *[amount of time, if applicable]*; you may have _____ attempts. When you are ready to perform this job, ask your instructor to observe the procedure and complete this form. To show mastery of this job, you must perform all steps to receive an overall competency rating of at least 2.

> **Competency Rating Scale**
>
> **3 — Skilled** — Meets all evaluation criteria and standards; performs task independently on first attempt; requires no additional practice or training.
>
> **2 — Moderately skilled** — Meets all evaluation criteria and standards; performs task independently; additional practice is recommended.
>
> **1 — Unskilled** — Is unable to perform the task; additional training required.
>
> ☒ **— Unassigned** — Job sheet task is not required or has not been performed.
>
> ✔ **Evaluator's Note:** Formulate and inform the candidate of the standards for this task (time allowed and number of attempts). Observe the candidate perform the task, check the step/key point under the appropriate attempt number as accomplished, record total time (if appropriate), and then use the rating scale above to assign an overall competency rating. If the candidate is unable to perform any step of this job, have the candidate review the materials and try again.

Introduction	To overcome friction loss and deliver large volumes of water over a long distance, two or more hoselines can be laid parallel and then "siamesed" into one line. This job sheet outlines the procedure for using the siamese fitting. The siamese adapter has two female connections and one male connection, whereas the wye adapter has two male connections and one female connection. This job sheet is written for a siamese appliance, but if you are alert, you should be able to perform the same evolution using a straight wye, double-male, and double-female fittings.
Equipment and Personnel	• One firefighter in protective clothing • Three sections of 2½-inch *(65 mm)* hose • Siamese appliance

Job Steps	Key Points	1	2	3
1. Carry the hose sections from the apparatus to the work area.	1. Using an approved unload and carry per departmental SOPs	___	___	___
2. Lay out two sections of 2½-inch *(65 mm)* hose.	2. a. Parallel	___	___	___
	b. Female ends side by side	___	___	___
	c. Flat	___	___	___
	d. Straight	___	___	___
3. Connect the siamese to the male coupling of one supply line.	3. a. First checking condition of threads and gaskets	___	___	___
	b. Using foot-tilt method	___	___	___
	c. Until hand-tight	___	___	___
4. Connect the siamese to the male coupling of the second supply line.	4. a. Using foot-tilt method	___	___	___
	b. Keeping first hoseline out of way	___	___	___
	c. Until hand-tight	___	___	___
5. Lay out a third section of 2½-inch *(65 mm)* hose.	5. a. Flat	___	___	___
	b. Straight	___	___	___
	c. Female end toward siamese	___	___	___
6. Connect the siamese to the female coupling of the third (single-discharge) line.	6. a. Using foot-tilt method	___	___	___
	b. Keeping supply lines between arms and keeping body over knees	___	___	___
	c. Turning female coupling clockwise with free hand	___	___	___
	d. Until hand-tight	___	___	___
7. Uncouple the siamese from the hoses.	7. a. Turning counterclockwise	___	___	___
	b. Using knee-press or stiff-arm method as necessary	___	___	___
8. Roll hose and return it and the siamese to storage.	8. Per departmental SOPs	___	___	___

Time (Total)	___	___	___

Evaluator's Comments _____

_____

_____

Job Sheet 12-6
Wye One Line into Two or More Lines

Name _____ Date _____

Evaluator _____ Overall Competency Rating _____

References	NFPA 1001, Fireground Operations 4-3.1, 4-3.2, 4-3.2a, 4-3.3 *Essentials*, pages 406, 407
Prerequisites	Firefighter I competency in coupling and uncoupling hose, particularly methods of uncoupling tight connections
Student's Instructions	To meet evaluation standards, you must perform this job within _____ *[amount of time, if applicable]*; you may have _____ attempts. When you are ready to perform this job, ask your instructor to observe the procedure and complete this form. To show mastery of this job, you must perform all steps to receive an overall competency rating of at least 2.

> **Competency Rating Scale**
>
> **3 — Skilled** — Meets all evaluation criteria and standards; performs task independently on first attempt; requires no additional practice or training.
>
> **2 — Moderately skilled** — Meets all evaluation criteria and standards; performs task independently; additional practice is recommended.
>
> **1 — Unskilled** — Is unable to perform the task; additional training required.
>
> ☒ — **Unassigned** — Job sheet task is not required or has not been performed.
>
> ✔ **Evaluator's Note:** Formulate and inform the candidate of the standards for this task (time allowed and number of attempts). Observe the candidate perform the task, check the step/key point under the appropriate attempt number as accomplished, record total time (if appropriate), and then use the rating scale above to assign an overall competency rating. If the candidate is unable to perform any step of this job, have the candidate review the materials and try again.

Introduction	Occasionally it is desirable to divide a line into two or more hoselines on the fireground. Various types of wye connections are available that the firefighter can use to do this — the most common being the 2½- to 1½-inch *(65 mm to 38 mm)* gated wye. This appliance is used to divide the 2½-inch *(65 mm)* line into two more easily handled 1½-inch *(38 mm)* lines.
Equipment and Personnel	• One firefighter in protective clothing • Sufficient 2½-inch and 1½-inch *(65 mm and 38 mm)* hose • 2½- to 1½-inch *(65 mm to 38 mm)* gated wye

Job Steps	Key Points	Attempt No. 1	2	3
1. Lay out the 2½-inch *(65 mm)* hoseline.	1. a. Flat	___	___	___
	b. Straight	___	___	___
2. Obtain and check gated wye.	2. a. Threads	___	___	___
	b. Gaskets	___	___	___
	c. Gates closed	___	___	___
	✔Note: Replace the gasket if it is worn. Replace the gated wye if the threads are damaged.			
3. Attach gated wye to male coupling of 2½-inch *(65 mm)* line.	3. a. Using foot-tilt method	___	___	___
	b. Turning clockwise	___	___	___
	c. Until hand tight	___	___	___
4. Lay out two 1½-inch *(38 mm)* hoselines.	4. a. Female ends facing gated wye	___	___	___
	b. Flat	___	___	___
	c. Straight	___	___	___
	d. Parallel	___	___	___
5. Attach 1½-inch *(38 mm)* lines to male outlets of gated wye.	5. a. Using foot-tilt method	___	___	___
	b. Turning clockwise	___	___	___
	c. Until hand tight	___	___	___
6. Check position of gate valves.	6. Closed	___	___	___
7. Uncouple hoses from gate valve.	7. a. Turning counterclockwise	___	___	___
	b. Using knee-press or stiff-arm method as necessary	___	___	___
8. Roll hose and return it and gated wye to storage.	8. Per departmental SOPs	___	___	___
	Time (Total)	___	___	___

Evaluator's Comments _____

_____

_____

Job Sheet 12-7
Service Test Hose

Name _____ Date _____

Evaluator _____ Overall Competency Rating _____

References	NFPA 1001, Prevention, Preparedness, and Maintenance 4-5.3, 4-5.3b *Essentials*, pages 437, 438, and 480–483
Prerequisites	Firefighter I hose competencies on coupling and uncoupling hose
Student's Instructions	To meet evaluation standards, you must perform this job within _____ *[amount of time, if applicable]*; you may have _____ attempts. When you are ready to perform this job, ask your instructor to observe the procedure and complete this form. To show mastery of this job, you must perform all steps to receive an overall competency rating of at least 2.

Competency Rating Scale

3 — Skilled — Meets all evaluation criteria and standards; performs task independently on first attempt; requires no additional practice or training.

2 — Moderately skilled — Meets all evaluation criteria and standards; performs task independently; additional practice is recommended.

1 — Unskilled — Is unable to perform the task; additional training required.

☒ — **Unassigned** — Job sheet task is not required or has not been performed.

✔ **Evaluator's Note:** Formulate and inform the candidate of the standards for this task (time allowed and number of attempts). Observe the candidate perform the task, check the step/key point under the appropriate attempt number as accomplished, record total time (if appropriate), and then use the rating scale above to assign an overall competency rating. If the candidate is unable to perform any step of this job, have the candidate review the materials and try again.

Introduction	Service testing is performed on in-service fire department hose to ensure that it can function under maximum pressure during fire fighting or other operations. Fire department hose should be tested annually, after being repaired, or after being run over by a vehicle. Unlined standpipe hose should be tested five years from the date of purchase, again at the eighth year, and every two years thereafter.

Before performing the service test, firefighters should inspect the hose for defects, coupling damage, and worn or defective gaskets. Any defects should be corrected if possible. If the damage is not repairable, the hose should be taken out of service.

The following procedure is for testing lined fire hose and large diameter hose. |

	Equipment and Personnel
Equipment and Personnel	• Two to four firefighters in protective clothing • Test length(s) of hose under 300 feet *(90 m)* • Fire department pumper • Nozzles with shut-off valves (appropriate number) • Hose test gate valves (appropriate number) • Spanner wrench • Rope hose tools or hose straps (one for each test length) • Hose gaskets (appropriate size) • Hose log • *NFPA 1962 Standard for the Care, Use, and Service Testing of Fire Hose Including Couplings and Nozzles* • Red tags for identifying failed sections of hose • White tags for identifying hose sections needing new couplings • Marker for marking each length with the test date • Chalk

Job Steps	Key Points	Attempt No. 1	2	3

1. Set up the hose testing area.	1. Area meets following criteria:
	  a. Isolated from traffic
	  b. Well-lighted
	  c. Smooth, clean surface
	  d. Slightly graded surface (if possible)
	  e. Accessible to water source
		___	___	___

2. Lay out the hose to be tested.	2. Straight runs	___	___	___
3. Connect a number of hose sections.	3. a. Checking gaskets before connecting	___	___	___
	b. Into test lengths of no more than 300 feet *(90 m)*	___	___	___
	c. Tightening couplings with spanner wrench	___	___	___
4. Connect hose test gate valves to each pumper discharge valve used.	4. a. All valves in OPEN position	___	___	___
	b. Counterclockwise until threads are set	___	___	___
	c. Clockwise until hand-tight	___	___	___
	d. Tightening each connection with spanner wrench	___	___	___
5. Connect the hose test length(s) to the test gate valve(s).	5. a. Counterclockwise until threads are set	___	___	___
	b. Clockwise until hand-tight	___	___	___
	c. Tightening each connection with spanner wrench	___	___	___

			Attempt No.		
Job Steps		**Key Points**	1	2	3

Job Steps	Key Points	1	2	3
6. Attach a rope hose tool or hose strap to each test length.	6. a. 10 to 15 inches *(250 mm to 375 mm)* from test valve connections	—	—	—
	b. Opposite end secured to discharge pipe or other nearby anchor	—	—	—
7. Attach a shutoff nozzle to the open end of each test length.	7. a. Counterclockwise until threads are set	—	—	—
	b. Clockwise until hand-tight	—	—	—
8. Open nozzle(s) for hose filling.	8. Holding above level of pump or hydrant discharge	—	—	—
9. Open discharge.	9. (None)	—	—	—
10. Fill each hoseline with water.	10. a. To hydrant pressure or to pump pressure of 50 psi *(350 kPa)*	—	—	—
	b. Permitting all air in hose to discharge	—	—	—
	c. Flowing discharged water *away* from test area	—	—	—
11. Close the nozzle(s).	11. a. After all air has been purged from test length(s)	—	—	—
	b. Slowly to prevent water hammer	—	—	—
12. Mark each hose jacket with chalk or pencil.	12. a. Against each coupling	—	—	—
	b. Hash mark on hose *and* coupling	—	—	—
13. Inspect the hoseline for kinks or twists.	13. Straightening any kinks or twists	—	—	—
14. Inspect each hose length for leaking couplings and leaks *behind* the couplings.	14. a. Retightening leaking coupling connections	—	—	—
	b. With spanner wrench	—	—	—
15. Remove failed lengths from the hoseline.	15. a. Relieving pressure	—	—	—
	b. Disconnecting couplings	—	—	—
	c. Red-tagging and removing from service lengths leaking from *behind* coupling	—	—	—

Job Steps	Key Points	Attempt No. 1 2 3
	d. Replacing gasket in lengths leaking at coupling, and starting over at Step 5	—— —— ——
16. Close each hose test gate valve.	16. (None)	—— —— ——
17. Increase the pump pressure.	17. To that required in *NFPA 1962 Standard for the Care, Use, and Service Testing of Fire Hose Including Couplings and Nozzles*	—— —— ——
18. Monitor the connections as pressure increases.	18. a. Checking for leakage	—— —— ——
	b. Standing to side, not over hose	—— —— ——
19. Maintain test pressure.	19. For 5 minutes	—— —— ——
20. Inspect all couplings during 5-minute pressurization.	20. a. Checking for leakage (weeping) at point of attachment	—— —— ——
	b. Standing to side, not over hose	—— —— ——
21. Reduce pump pressure.	21. a. After 5 minutes	—— —— ——
	b. Slowly	—— —— ——
22. Close each discharge valve.	22. When pump pressure at zero	—— —— ——
23. *(Driver / operator)* Disengage the pump.	23. (None)	—— —— ——
24. Open each nozzle.	24. a. Slowly	—— —— ——
	b. To bleed off pressure in test lengths	—— —— ——
	c. Discharging water *away* from test area	—— —— ——
25. Break all hose connections.	25. Draining water *away* from test area	—— —— ——
26. Observe the marks placed on the hose at the couplings.	26. To determine whether coupling has moved during test	—— —— ——

Job Steps	Key Points	Attempt No.		
		1	2	3
	✔ **Note:** Expect a ¹⁄₁₆- to ⅛-inch *(2 mm to 3 mm)* uniform movement of the coupling on newly coupled hose. This slippage is normal during initial testing, but should not occur during subsequent tests.			
27. Tag hose section(s) as necessary.	27. Red tag for those that leaked or failed in any way	___	___	___
28. Record your test results.	28. a. In hose logbook	___	___	___
	b. For each section of hose	___	___	___
	c. Dating each hose section	___	___	___
	Time (Total)	___	___	___

Evaluator's Comments _____

_____

_____

Chapter 12 Review Test

> **➡ Directions:** This review test covers the Firefighter II material in Chapter 12 of your ***Essentials of Fire Fighting*** text. It may be assigned as a study aid (self-test) or may be administered by your instructor as a pretest or posttest.
>
> When used as a study aid, try to answer the questions without referring to the page numbers in ***Essentials*** or your ***Firefighter II Student Applications*** workbook *(SA)* on which the answers can be found until after you have completed the entire test. Then check your answers against those on the pages provided in parentheses.
>
> When administered by your instructor as a pretest or posttest, read each of the test questions carefully. Choose the best response and then darken the corresponding letter on your answer sheet.
>
> This chapter review test contains 33 multiple-choice questions, each worth 3 points. To pass the test, you must achieve at least 84 of the 99 points possible.

1. Which of the following is a hose appliance? *(408)*

   A. Hose clamp

   B. Water thief

   C. Hose roller

   D. Hose jacket

2. Which of the following is a hose tool? *(410)*

   A. Siamese

   B. Reducer

   C. Wye

   D. Hose roller

3. What type of hose valve is illustrated below? *(407)*

   A. Ball valve

   C. Clapper valve

   B. Gate valve

   D. Butterfly valve

---

4. What type of hose valve is illustrated below? *(407)*

A. Gate valve
B. Butterfly valve

C. Clapper valve
D. Ball valve

5. Which valve below fits the following description: *A flat disk hinged on one side to swing like a door; used to allow only one intake hose to be connected and charged before the addition of more hoses? (406)*

A. Clapper valve
B. Butterfly valve

C. Gate valve
D. Ball valve

6. What hose appliance is illustrated below? *(407)*

A. Wye
B. LDH manifold
C. Water thief
D. Siamese

7. What type of valves allow only one intake hose to be connected and charged before the addition of more hoses? *(407, 408)*

A. Ball valve
B. Gate Valve
C. Clapper valve
D. Butterfly valve

8. For which of the following would a wye valve device be most appropriate? *(406, 407)*

A. Extending and dividing a single attack line for a two-pronged fire attack with two 1½-inch *(38 mm)* or two 2½-inch *(65 mm)* lines

B. Feeding LDH when multiple smaller hoselines have to be used in the same relay

C. Overcoming friction loss in hose lays that need to carry a large flow or cover a long distance

D. Supplying ladder pipes not equipped with a permanent waterway

9. What type of hose appliance is illustrated below? *(408)*

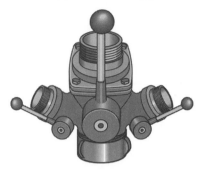

    A. LDH manifold   C. Water thief

    B. Siamese   D. Wye

10. Firefighter A says that a siamese connection brings two or more hoselines into one hoseline or device.

    Firefighter B says that a wye connection divides a hoseline into two or more hoselines. Who is right? *(406–408)*

    A. Firefighter A   C. Both A and B

    B. Firefighter B   D. Neither A nor B

11. With what type of lay are hydrant valves used? *(408)*

    A. Split lay   C. Combination lay

    B. Forward lay   D. Reverse lay

12. What type of hose fitting is illustrated below? *(410)*

    A. Reducer   C. Double female adapter

    B. Double male adapter   D. Elbow

13. What type of hose fitting is illustrated below? *(409)*

    A. Double female adapter

    B. Double male adapter

    C. Reducer

    D. Elbow

14. What type of hose fitting is illustrated below? *(409)*

   A. Double female adapter   C. Double male adapter

   B. Reducer   D. Elbow

15. Firefighter A says that extending a line with a reducer allows the option of adding another line if needed.

   Firefighter B says that a double female and male adapters are most commonly used when a pumper that is set up for a forward lay is used for a reverse lay.

   Who is right? *(409)*

   A. Firefighter A   C. Both A and B

   B. Firefighter B   D. Neither A nor B

16. What is the purpose of an elbow fitting? *(410)*

   A. Connecting a smaller hoseline to a larger one

   B. Changing the direction of flow

   C. Closing off male couplings

   D. Closing off female couplings

17. What is the purpose of a hose roller? *(410)*

   A. Stopping water flow through a line

   B. Protecting the hose from sharp edges

   C. Removing water from washed hose

   D. Storing booster line and LDH hose

18. What hose tool is illustrated below? *(411)*

   A. Hose clamp

   B. Chafing block

   C. Hose jacket

   D. Hose roller

19. What type of hose clamp is illustrated below? *(411)*

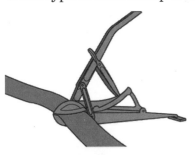

   A. Screw-down

   B. Press-down

   C. Pull up

   D. Hydraulic

20. When using a hose clamp, approximately how far behind the apparatus should it be applied? *(411)*

   A. 5 feet *(1.5 m)*

   B. 20 feet *(6 m)*

   C. 12 feet *(4 m)*

   D. 15 feet *(5 m)*

21. Apply a hose clamp approximately ____ from the coupling on the incoming water side. *(411)*

   A. 20 feet *(6 m)*   C. 15 feet *(5 m)*

   B. 5 feet *(1.5 m)*   D. 12 feet *(4 m)*

22. Why should a hose clamp be closed slowly? *(412)*

   A. To prevent the handle from snapping open

   B. To prevent the clamp from pinching the hose

   C. To avoid water hammer

   D. To secure the locking device

23. Firefighter A says that hard-suction hose strainers may be attached to either end of a hard-suction sleeve.

   Firefighter B says hard-suction hose strainers should not be allowed to rest on the bottom of a swimming pool. *(410)*

   Who is right?

   A. Firefighter A   C. Both A and B

   B. Firefighter B   D. Neither A nor B

24. What is the primary purpose of a spanner wrench? *(412)*

   A. Prying couplings apart

   B. Hammering coupling lugs to loosen or tighten

   C. Opening and closing hose valves

   D. Loosening and tightening couplings

25. What is the primary use of a rubber mallet? *(412)*

    A. Hammering intake hose coupling lugs to loosen or tighten

    B. Straightening slightly bent couplings and lugs

    C. Seating gaskets

    D. Operating wye valve levers

26. Where are chafing blocks particularly useful? *(413)*

    A. At corners and obstructions when advancing a line into a building

    B. Between the intake hose and the pavement or curb

    C. On the edges of windowsills and parapets when hoisting hose

    D. At the front of the hose bed where hose folds contact the bed

27. Which of the following is **not** a use for hose rope, hose strap, and hose chain tools? *(413)*

    A. Securing firefighters to ladders

    B. Securing hose to ladders and other fixed objects

    C. Carrying and pulling fire hose

    D. Securing a pressurized hose when applying water

28. How often should fire hose be service tested? *(437)*

    A. Once every 6 months, after being repaired, and after being run over by a vehicle

    B. Once every 5 years, after being repaired, and after being run over by a vehicle

    C. Once a year, after being repaired, and after being run over by a vehicle

    D. Once every 4 months, after being repaired, and after being run over by a vehicle

29. Firefighter A says that acceptance testing of fire hose is more rigorous than service testing.
    Firefighter B says that acceptance testing of fire hose is performed by the manufacturer.
    Who is right? *(437)*

    A. Firefighter A   C. Both A and B

    B. Firefighter B   D. Neither A nor B

30. What is the purpose of a hose test gate valve? *(437)*

    A. To measure pressure in the line

    B. To meter the amount of water flowing through the hoseline

    C. To allow for the line to be pressure tested when wyed

    D. To prevent water from surging through the hose if it fails

31. When being service tested, why should hose test lengths not exceed 300 feet *(90 m)*? *(437)*

    A. Longer lengths lose too much pressure to friction loss.

    B. Longer lengths cannot be accommodated in the hose testing machine.

    C. Longer lengths provide unreliable readings.

    D. Longer lengths are more difficult to purge of air.

32. When service testing hose, why should the testing area be kept free of water? *(438)*

    A. To aid in detecting minor leaks around the couplings

    B. To prevent the hose jacket from expanding slightly

    C. To prevent the hose test gate valve from malfunctioning

    D. To aid the firefighter in reading the hose test markings

33. Firefighter A says that when service testing hose, the hose should be connected to discharges on the pump panel side of the apparatus.

    Firefighter B says that when service testing hose, all personnel operating in the area of the pressurized hose should wear at least goggles as a safety precaution.

    Who is right? *(437)*

    A. Firefighter A

    B. Firefighter B

    C. Both A and B

    D. Neither A nor B

REVIEW TEST ANSWER SHEET

	A	B	C	D
1.	○	○	○	○
2.	○	○	○	○
3.	○	○	○	○
4.	○	○	○	○
5.	○	○	○	○
6.	○	○	○	○
7.	○	○	○	○
8.	○	○	○	○
9.	○	○	○	○
10.	○	○	○	○
11.	○	○	○	○
12.	○	○	○	○
13.	○	○	○	○
14.	○	○	○	○
15.	○	○	○	○
16.	○	○	○	○
17.	○	○	○	○
18.	○	○	○	○
19.	○	○	○	○
20.	○	○	○	○
21.	○	○	○	○
22.	○	○	○	○
23.	○	○	○	○
24.	○	○	○	○
25.	○	○	○	○
26.	○	○	○	○
27.	○	○	○	○
28.	○	○	○	○
29.	○	○	○	○
30.	○	○	○	○
31.	○	○	○	○
32.	○	○	○	○
33.	○	○	○	○

	A	B	C	D
34.	○	○	○	○
35.	○	○	○	○
36.	○	○	○	○
37.	○	○	○	○
38.	○	○	○	○
39.	○	○	○	○
40.	○	○	○	○
41.	○	○	○	○
42.	○	○	○	○
43.	○	○	○	○
44.	○	○	○	○
45.	○	○	○	○
46.	○	○	○	○
47.	○	○	○	○
48.	○	○	○	○
49.	○	○	○	○
50.	○	○	○	○
51.	○	○	○	○
52.	○	○	○	○
53.	○	○	○	○
54.	○	○	○	○
55.	○	○	○	○
56.	○	○	○	○
57.	○	○	○	○
58.	○	○	○	○
59.	○	○	○	○
60.	○	○	○	○
61.	○	○	○	○
62.	○	○	○	○
63.	○	○	○	○
64.	○	○	○	○
65.	○	○	○	○
66.	○	○	○	○
67.	○	○	○	○

	A	B	C	D
68.	○	○	○	○
69.	○	○	○	○
70.	○	○	○	○
71.	○	○	○	○
72.	○	○	○	○
73.	○	○	○	○
74.	○	○	○	○
75.	○	○	○	○
76.	○	○	○	○
77.	○	○	○	○
78.	○	○	○	○
79.	○	○	○	○
80.	○	○	○	○
81.	○	○	○	○
82.	○	○	○	○
83.	○	○	○	○
84.	○	○	○	○
85.	○	○	○	○
86.	○	○	○	○
87.	○	○	○	○
88.	○	○	○	○
89.	○	○	○	○
90.	○	○	○	○
91.	○	○	○	○
92.	○	○	○	○
93.	○	○	○	○
94.	○	○	○	○
95.	○	○	○	○
96.	○	○	○	○
97.	○	○	○	○
98.	○	○	○	○
99.	○	○	○	○
100.	○	○	○	○

Name _____

Date _____
Score _____

Chapter 12 Competency Profile

Student Name _____ Soc. Sec. No. _____

Last   First   Middle

Fire Department _____

Address _____

Phone _____

Home Address _____

Phone _____

Date of Enrollment _____ - _____ - _____   Total Class Hours _____

Date of Withdrawal _____ - _____ - _____   Total Hours Absent _____

Date of Completion _____ - _____ - _____

Instructor's Name _____   Session Dates _____

Instructor's Directions

1. Check the candidate's competency rating (3, 2, 1, ☒) for each performance test task and psychomotor lesson objective (practical activity and job sheets) listed below.

2. List any additional performance tasks or psychomotor objectives (job sheets or practical activity sheets) under "Other," and check the candidate's competency rating.

3. Record the candidate's cognitive scores (written lesson tests and *administered* chapter review tests) in the spaces provided.

Level				Psychomotor Competencies
3	**2**	**1**	**☒**	

Practical Activity Sheets

☐	☐	☐	☐	PAS 12-1 — Select Adapters and Appliances for Given Fireground Situations
☐	☐	☐	☐	Other _____
☐	☐	☐	☐	_____

Job Sheets

☐	☐	☐	☐	JS 12-1 — Apply a Hose Jacket to a Leaking Coupling
☐	☐	☐	☐	JS 12-2 — Replace a Section of Hose in a Charged Line
☐	☐	☐	☐	JS 12-3 — Extend a Charged Hoseline
☐	☐	☐	☐	JS 12-4 — Reduce a Charged Hoseline
☐	☐	☐	☐	JS 12-5 — Siamese Two or More Lines into One Line
☐	☐	☐	☐	JS 12-6 — Wye One Line into Two or More Lines

☐	☐	☐	☐
☐	☐	☐	☐
☐	☐	☐	☐

Psychomotor Competencies

JS 12-7 — Service Test Hose

Other _____

_____

Chapter 12 Performance Test

☐	☐	☐	☐

Task 1 — Perform an annual service test on fire hose — given a pump, a marking device, pressure gauges, a timer, record sheets, and related equipment

☐	☐	☐	☐

Task 2 — Select the appropriate tool and/or appliance, and correct two of the following fireground problems:

☐ Hose couplings on an uncharged hose are loose.

☐ Fire hydrant needs to be opened.

☐ Hose must be laid where vehicles must cross it.

☐ Hose is being damaged by rubbing on curb.

☐ Leaking hose couplings on charged hose cannot be further tightened.

☐ A section of hose must be replaced in a charged hoseline.

☐ A charged hoseline is too short.

☐ Hoseline flow is too great for salvage and overhaul.

☐ Friction loss is reducing flow, and water must be delivered over a long distance.

☐ Hoseline must be divided into two or more lines.

☐	☐	☐	☐
☐	☐	☐	☒

Other _____

_____

Points Achieved	Points Needed/ Total

Cognitive Competencies

Written Test

Points Achieved	Points Needed/Total	
_____	10/12	1. Identify types of valves and valve devices.
_____	6/8	2. Match types of valves to their functions.
_____	12/14	3. Identify hose fitting appliances.
_____	20/24	4. Identify tools used with hose.
_____	18/20	5. Match hose appliances and tools to their uses in specific fireground situations.
		6. Evaluated on Practical Activity Sheet 12-1

Points Achieved	Points Needed/ Total	Cognitive Competencies
		7. Evaluated on Job Sheets 12-1 through 12-6
_____	10/12	8. Select facts about service testing hose.
_____	6/6	9. List safety guidelines for service testing hose.
		10. Evaluated on Job Sheet 12-7

Review Test

_____   Chapter 12 Review Test

Instructor's Signature _____ **Date** _____

Student's Signature _____ **Date** _____

STUDENT APPLICATIONS

FOURTH EDITION

ESSENTIALS OF FIRE FIGHTING

LESSON 13

FOAM FIRE STREAMS

FIREFIGHTER II

FIRE PROTECTION PUBLICATIONS
OKLAHOMA STATE UNIVERSITY

Study Objectives

LESSON OBJECTIVE

After completing this lesson, you will be able to mix foam concentrate and assemble and operate a foam fire stream system.

ENABLING OBJECTIVES

After reading Chapter 13 of *Essentials,* pages 498 through 517, and completing related activities, you will be able to —

1. Describe the basic methods by which foam prevents or controls a hazard.

2. Classify flammable liquids as hydrocarbon or polar solvent fuels.

3. Explain how foam is generated.

4. Describe the components of foam production.

5. List factors that affect foam expansion.

6. Classify foams by their expansion ratios.

7. Distinguish between characteristics of Class A and Class B foams

8. List factors that affect Class B foam application rates.

9. Select facts about proportioning.

10. Match methods of proportioning to their descriptions.

11. Select facts about proportioners.

12. **Select foams for specific fire situations. (*Practical Activity Sheet 13-1*)**

13. Match types of handline foam nozzles to their uses.

14. **Select nozzles for specific fire situations. (*Practical Activity Sheet 13-2*)**

15. List reasons for poor foam generation.

16. Match foam application methods to their uses.

17. List types of hazards associated with foam use.

18. **Install an in-line foam eductor and operate a high-expansion foam generator. (*Job Sheet 13-1*)**

Study Sheet

Introduction

This study sheet is intended to help you learn the Firefighter II material in Chapter 13 of *Essentials of Fire Fighting*, Fourth Edition. You may use it for self-study, or you may use it to review material that will be covered in the lesson and chapter review tests. The numbers in parentheses are the pages in *Essentials* on which the answers or terms can be found.

Chapter Vocabulary

Be sure you know the chapter-related meanings of the following terms and abbreviations. Use the glossary in *Fire Service Orientation and Terminology* or a dictionary if you cannot determine the meaning of the term from its context.

- Aerated *(499)*
- AFFF *vs.* FFFP *(504)*
- Batch-mixing *(508)*
- Biodegradable *(515)*
- Cooling *(498)*
- Eductor *(508)*
- Foam concentrate *vs.* foam solution *(499)*
- Foam proportioner *(499)*

- Induction *(508)*
- Injection *(508)*
- Premixing *(508, 509)*
- Proportioned *(499)*
- Separating *(498)*
- Suppressing *(498)*
- Venturi principle *(510)*

Study Questions & Activities

1. In general, how does fire fighting foam work? *(498)*

   _____

   _____

   _____

   _____

2. Describe the three ways that foam extinguishes fire. *(498)*

   a. _____

   b. _____

   c. _____

3. Under what circumstances is foam more effective than water as an extinguishing agent? *(498)*

   _____

   _____

   _____

   _____

4. Because foams in use today are of the mechanical type, what must be done before they can be used? *(499)*

_____

_____

_____

5. What four elements are necessary to produce high-quality fire fighting foam? *(498)*

   a. _____

   b. _____

   c. _____

   d. _____

6. Describe hydrocarbon and polar solvent fuels. *(498, 499)*

   a. Hydrocarbon fuels _____

   b. Polar solvent fuels_____

7. Provide specific examples of hydrocarbon and polar solvent fuels. *(498, 499)*

   Hydrocarbon fuels   Polar solvent fuels

   a. _____   a. _____

   b. _____   b. _____

   c. _____   c. _____

   d. _____   d. _____

8. Define *foam expansion. (499)*

   _____

9. What four factors determine a foam solution's degree of expansion when it is aerated? *(500)*

   a. _____

   b. _____

   c. _____

   d. _____

10. Distinguish among low-, medium-, and high-expansion foams. *(500)*

    a. Low-expansion foam _____

    _____

    b. Medium-expansion foam _____

    _____

    c. High-expansion foam _____

    _____

11. Why is it important to identify the type of fuel involved before applying foam? *(500)*

    _____

    _____

12. With what nozzles, devices, and systems may Class A foams be used? *(500)*

_____

_____

13. What variables affect the rate of application for Class B foams? *(504)*

_____

_____

_____

_____

_____

14. Why is there a difference between foam application rates for ignited fuels and those for unignited fuel spills? *(505)*

_____

_____

_____

_____

15. Explain how the proportioning of Class A foams can be adjusted to achieve specific objectives. *(505)*

_____

_____

_____

_____

_____

16. Briefly explain the four basic methods by which foam may be proportioned. *(508, 509)*

   a. Induction _____

_____

_____

_____

_____

   b. Injection _____

_____

_____

   c. Batch mixing _____

_____

_____

   d. Premixing _____

_____

_____

17. What is the difference between an in-line eductor and a foam nozzle eductor? *(510)*

_____

_____

_____

_____

18. What two basic pieces of equipment are required to produce a foam fire stream — along with a pump to supply water and hose to transport the foam? *(509)*

   a. _____

   b. _____

19. What are the two basic principles by which foam proportioning devices work? *(509)*

   a. _____

   _____

   b. _____

   _____

20. What is the most common type of portable foam proportioner? *(510)*

_____

21. What is the maximum height above the concentrate level for positioning an in-line eductor? *(510)*

_____

22. List three disadvantages of foam nozzle eductors. *(510)*

   a. _____

   b. _____

   c. _____

23. Name the three types of apparatus-mounted proportioners. *(511)*

   a. _____

   _____

   b. _____

   _____

   c. _____

   _____

24. Distinguish among the following types of foam delivery devices. *(511–513)*

   a. Solid bore nozzle _____

   _____

   _____

   b. Fog nozzle _____

   _____

   _____

c. Air-aspirating nozzle _____

_____ SA 13 —

d. Water-aspirating type nozzle _____

_____

_____

_____

e. Mechanical blower generator _____

_____

_____

_____

_____

25. List the eight most common reasons for failure to generate foam when operating a foam system. *(513)*

a. _____

b. _____

c. _____

d. _____

e. _____

f. _____

g. _____

h. _____

26. Briefly describe the foam application techniques listed below. *(514)*

a. Roll-on method _____

_____

_____

b. Bank-down method _____

_____

_____

c. Rain-down method _____

_____

_____

27. Describe health hazards and equipment hazards associated with foams. *(515)*

a. Health hazards _____

_____

_____

b. Equipment hazards _____

_____

_____

28. Define the biodegradability of foam, and explain how it can damage waterways. *(515)*

_____

_____

_____

_____

Practical Activity Sheet 13-1
Select Foam for Specific Fire Situations

Name _____ Date _____

Evaluator _____ Overall Competency Rating _____

References	NFPA 1001, Fireground Operations, 4-3.1 *Essentials*, Table 13.1, pages 501–504
Prerequisite	None
Introduction	With the increased use and transportation of flammable and combustible liquids, incidents involving these substances are becoming more common. However, as with any fire fighting technique, the foam and application method used must be appropriate for the situation. If the incorrect foam is used, the attack may be ineffective and, beyond delaying extinguishment, may even worsen the situation. With this knowledge, you will be able to select the correct type of foam for specific fire situations.
Directions	Read each situation described. Refer to Table 13.1 in *Essentials* or other resources provided by your instructor to determine the type of foam that should be used. Write your answers on the blanks.
Activity	SITUATION 1: A fire in an underground tank containing fuel oil _____ SITUATION 2: A basement fire in a single-family wood-frame residence involving newspapers and the house structural materials _____ SITUATION 3: A wildland fire _____ SITUATION 4: A rupture of an oil tanker at sea _____ SITUATION 5: A fire occurring in a collision between a gasoline bulk transport vehicle and a flat-bed truck loaded with large barrels of lacquer thinner _____

Competency Rating Scale

3 — Skilled — All 5 situations answered appropriately per suggested answers in *Instructor's Guide*; student requires no additional practice.

2 — Moderately skilled — At least 3 of the 5 situations answered appropriately per suggested answers in *Instructor's Guide*; student may benefit from additional practice.

1 — Unskilled — Fewer than 3 of the 5 situations answered appropriately per suggested answers in *Instructor's Guide*; student requires additional practice and reevaluation.

☒ — Unassigned — Task is not required or has not been performed.

✔**Evaluator's Note:** Evaluate the results as indicated below. Assign an overall competency level, and record this competency rating on both the student's practical activity sheet and competency profile.

To show competency in this objective, the student must achieve an overall rating of at least 2.

Answers	Correct	Incorrect
All answers evaluated per answers in ***Instructor's Guide.***		
SITUATION 1	☐	☐
SITUATION 2	☐	☐
SITUATION 3	☐	☐
SITUATION 4	☐	☐
SITUATION 5	☐	☐

Practical Activity Sheet 13-2
Select Nozzles for Specific Fire Situations

Name _____ Date _____

Evaluator _____ Overall Competency Rating _____

References	NFPA 1001, Fireground Operations, 4-3.1 *Essentials*, 510–513 and Table 13.1, pages 501–504
Prerequisite	None
Introduction	For effective extinguishment and control of fire and spills using foam, the correct proportioner and delivery device must be used. If incorrect equipment is used, the attack may be ineffective and, beyond delaying extinguishment, may even worsen the situation. With this knowledge, you will be able to select the correct foam nozzles for specific fire situations and attack methods.
Directions	Read each situation described and determine the type of nozzle that should be used. Write your answers on the blanks.
Activity	**SITUATION 1:** An unignited spill of gasoline attacked with fluoroprotein foam _____ **SITUATION 2:** A kerosene leak from a tank car; the leak is to be suppressed with alcohol-resistant AFFF, and the objective is to cool the container and suppress vapors during a prolonged response at a derailment involving many tank cars _____ **SITUATION 3:** A widespread wildland fire on which Class A foam is being applied from apparatus _____ **SITUATION 4:** A fire in a coal mine _____ **SITUATION 5:** An unignited spill involving crude petroleum that is being attacked with AFFF foam prepared as a medium-expansion foam _____

Competency Rating Scale

3 — Skilled — All 5 situations answered appropriately per suggested answers in *Instructor's Guide*; student requires no additional practice.

2 — Moderately skilled — At least 3 of the 5 situations answered appropriately per suggested answers in *Instructor's Guide*; student may benefit from additional practice.

1 — Unskilled — Fewer than 3 of the 5 situations answered appropriately per suggested answers in *Instructor's Guide*; student requires additional practice and reevaluation.

☒ **— Unassigned** — Task is not required or has not been performed.

✔**Evaluator's Note:** Evaluate the results as indicated below. Assign an overall competency level, and record this competency rating on both the student's practical activity sheet and competency profile.

To show competency in this objective, the student must achieve an overall rating of at least 2.

Answers	Correct	Incorrect
All answers evaluated per answers in ***Instructor's Guide.***		
SITUATION 1	☐	☐
SITUATION 2	☐	☐
SITUATION 3	☐	☐
SITUATION 4	☐	☐
SITUATION 5	☐	☐

Job Sheet 13-1
Install an In-Line Foam Eductor and Operate a High-Expansion Foam Generator

Name _____ Date _____

Evaluator _____ Overall Competency Rating _____

References	NFPA 1001, Fireground Operations 4-3.1 *Essentials*, pages 510, 513, 516, 517
Prerequisites	None
Student's Instructions	To meet evaluation standards, you must perform this job within _____ *[amount of time, if applicable]*; you may have _____ attempts. When you are ready to perform this job, ask your instructor to observe the procedure and complete this form. To show mastery of this job, you must perform all steps to receive an overall competency rating of at least 2.

> **Competency Rating Scale**
>
> **3 — Skilled** — Meets all evaluation criteria and standards; performs task independently on first attempt; requires no additional practice or training.
>
> **2 — Moderately skilled** — Meets all evaluation criteria and standards; performs task independently; additional practice is recommended.
>
> **1 — Unskilled** — Is unable to perform the task; additional training required.
>
> **☒ — Unassigned** — Job sheet task is not required or has not been performed.
>
> ✔ **Evaluator's Note:** Formulate and inform the candidate of the standards for this task (time allowed and number of attempts). Observe the candidate perform the task, check the step/key point under the appropriate attempt number as accomplished, record total time (if appropriate), and then use the rating scale above to assign an overall competency rating. If the candidate is unable to perform any step of this job, have the candidate review the materials and try again.

Introduction	To provide a foam stream, the firefighter or apparatus driver must be able to correctly assemble the components of the system. The following procedure describes the steps for placing a foam line in service using an in-line proportioner.
Equipment and Personnel	• One driver/operator • One pumper • One firefighter in protective clothing • Foam eductor • Hose and nozzle compatible with eductor • Two buckets of foam concentrate • Water supply

Job Steps	Key Points	Attempt No. 1	2	3
1. Select the proper foam concentrate.	1. a. Appropriate for fuel type	___	___	___
	b. Using Table 13-1 in *Essentials* if necessary	___	___	___
2. Check the eductor and nozzle.	2. Hydraulically compatible (rated for the same flow)	___	___	___
3. Check the foam concentration listing on the side of the container.	3. Matches eductor percentage rating	___	___	___
4. Set the eductor to the proper rating if adjustable.	4. Same as foam concentration listing on side of container	___	___	___
5. Attach the eductor to a hose or discharge outlet.	5. a. Avoiding kinks in hose	___	___	___
	b. Avoiding connections to discharge elbows	___	___	___
	c. Making sure that ball gates are completely open if eductor is connected directly to pump discharge outlet	___	___	___
6. Select and attach the attack hoseline.	6. a. To desired nozzle	___	___	___
	b. To discharge end of eductor	___	___	___
	c. Length of the hose not exceeding eductor manufacturer's recommendations	___	___	___
7. Open containers of foam concentrate.	7. Enough estimated to handle task without interruption in flow of concentrate	___	___	___
8. Position the containers of concentrate.	8. a. Near eductor	___	___	___
	b. Bottom of containers no more than 6 feet *(2 m)* below eductor	___	___	___
9. Place the eductor siphon tube into the concentrate.	9. (None)	___	___	___
10. Increase the water supply pressure.	10. To that recommended by eductor manufacturer	___	___	___
11. Signal the driver/operator.	11. To charge the attack line	___	___	___

Job Steps	Key Points	Attempt No.		
		1	2	3
12. Apply the foam.	12. According to manufacturer's directions	—	—	—
	Time (Total)	—	—	—

Evaluator's Comments _____

_____

_____

Chapter 13 Review Test

→**Directions:** This review test covers the Firefighter II material in Chapter 13 of your ***Essentials of Fire Fighting*** text. It may be assigned as a study aid (self-test) or may be administered by your instructor as a pretest or posttest.

When used as a study aid, try to answer the questions without referring to the page numbers in ***Essentials*** or your ***Firefighter II Student Applications*** workbook *(SA)* on which the answers can be found until after you have completed the entire test. Then check your answers against those on the pages provided in parentheses.

When administered by your instructor as a pretest or posttest, read each of the test questions carefully. Choose the best response and then darken the corresponding letter on your answer sheet.

This chapter review test contains 50 multiple-choice questions, each worth 2 points. To pass the test, you must achieve at least 84 of the 100 points possible.

1. In what way is foam an effective fire fighting agent? *(498)*

   A. By forming a blanket on the burning fuel

   B. By releasing water onto the fire as the foam breaks down

   C. By preventing the release of flammable vapors

   D. All of the above

2. What are the two types of flammable liquid fuels for which foam is a particularly effective fire fighting agent? *(498)*

   A. Hydrocarbon fuels and crude oil fuels

   B. Polar solvent fuels and ketone fuels

   C. Hydrocarbon fuels and polar solvent fuels

   D. Hydrocarbon solvent fuels and polar fuels

3. What term is applied to a foam's ability to create a barrier between a fire and the fuel? *(498)*

   A. Separating   C. Suppressing

   B. Cooling   D. Smothering

4. Firefighter A says that gasoline, naphtha, jet fuel, and kerosene are examples of hydrocarbon fuels.

   Firefighter B says that alcohol, acetone, lacquer thinner, and esters are examples of hydrocarbon fuels.

   Who is right? *(498, 499)*

   A. Firefighter A   C. Both A and B

   B. Firefighter B   D. Neither A nor B

5. Which of the following is **not** a fire that a specialized foam would aid in extinguishing? (499)

    A. Deep-seated Class A fire

    B. Electrical equipment fire

    C. Confined- or enclosed-space fire

    D. Fire involving acid spills, pesticides, or other hazardous liquids

6. What term is used to describe the raw foam liquid prior to the introduction of water and air? (499)

    A. Foam proportioner   C. Foam concentrate

    B. Foam solution   D. Finished foam

7. What is the purpose for introducing air during the generation of fire fighting foams? (499)

    A. To reduce the cost of the foam

    B. To accelerate the rate of water absorption

    C. To improve the flow of the foam through the dispersion nozzle

    D. To produce uniform-sized bubbles to provide a longer lasting blanket

8. Which of the following is **not** a major factor in the amount of expansion in a finished foam? (500)

    A. Type of foam concentrate used

    B. Temperature of the air used for aeration

    C. Accurate proportioning of the foam concentrate in the solution

    D. Method of aspiration

9. According to *NFPA 11 Standard for Low-Expansion Foam*, a low-expansion foam is one that has an air/solution ratio of ___ parts finished foam for every part of foam solution. (500)

    A. Less than 10:1   C. 20:1 to 200:1

    B. Up to 20:1   D. 200:1 to 1000:1

10. Firefighter A says that the mixing of water with foam concentrate is called *proportioning*.

Firefighter B says that the dispersion of the foam from the end of the nozzle is called *aeration*.

Who is right? (499)

    A. Firefighter A   C. Both A and B

    B. Firefighter B   D. Neither A nor B

11. How do surfactants increase foam effectiveness? (500)

    A. Increasing the foam expansion rate

    B. Reducing surface tension to provide better water penetration

    C. Increasing acidity to heighten the ionic bond

    D. Providing greater cooling capacity through blanketing action

12. What does the abbreviation *CAFS* stand for? *(500)*

    A. Concentrated acidic foam system

    B. Compressed air foam system

    C. Class A foam solution

    D. Caustic activated foam solution

13. What are the two bases for Class B foam concentrates? *(500)*

    A. Hydrocarbon and polar

    B. Protein and polar

    C. Synthetic and hydrocarbon

    D. Protein and synthetic

14. Which of the following is an acceptable device for discharging Class A foam? *(500)*

    A. Fog nozzle

    B. Medium- and high-expansion devices

    C. Air-aspirating foam nozzle

    D. All of the above

15. What does the abbreviation *AFFF* stand for? *(504)*

    A. Aqueous film forming foam

    B. Acidic film forming fluoroprotein

    C. Advanced fluoroprotein forming foam

    D. Aspirated foam forming filter

16. What does the abbreviation *FFFP* stand for? *(504)*

    A. Film forming foam proportioner

    B. Foam forming fluoroprotein

    C. Film forming fluoroprotein foam

    D. Fluoroprotein foam filtering proportioner

17. Firefighter A says that Class B foams should be applied to spills at the same rate without regard to whether the fuel has ignited because application rates are based on the foam and not on the fuel.

    Firefighter B says that Class B foam application can be discontinued prior to extinguishment once the unignited portion of the spill has been totally blanketed.

    Who is right? *(505)*

    A. Firefighter A   C. Both A and B

    B. Firefighter B   D. Neither A nor B

18. On which class of fire is foam especially effective? *(498)*

    A. Class A   C. Class C

    B. Class B   D. Class D

19. Firefighter A says that salt water is **not** acceptable for mixing with most foam concentrates.

Firefighter B says that Class A foams can be proportioned over a range of percentages as defined by their manufacturers to achieve specific objectives.

Who is right? *(505)*

A. Firefighter A   C. Both A and B

B. Firefighter B   D. Neither A nor B

20. Use of a device with a restricted diameter to draw foam concentrate into a hoseline is an example of what method of foam proportioning? *(508)*

A. Premixing   C. Induction

B. Batch-mixing   D. Injection

21. Which of the following is **not** a disadvantage of batch-mixing? *(508)*

A. Batch-mixing may not be effective on large incidents in which the tank becomes empty.

B. Batch-mixing is difficult to accomplish and highly inaccurate.

C. When batch-mixing, some foams require that the concentrate and water be circulated for a period of time before discharge.

D. Batch-mixing should be used only as a last resort for Class B foams.

22. Which proportioning method uses a pump or head pressure to force concentrate into the fire stream at the correct ratio? *(508)*

A. Induction   C. Premixing

B. Injection   D. Batch-mixing

23. What method of proportioning is generally used with portable fire extinguishers? *(508, 509)*

A. Batch-mixing   C. Injection

B. Premixing   D. Induction

24. Class B foams are normally mixed in proportions of ___. *(507)*

A. 1% to 6%   C. 6% to 12%

B. 3% to 6%   D. 6% to 30%

25. What is another term commonly used to describe an induction foam proportioner? *(508)*

A. Eductor   C. In-line Venturi

B. Injector   D. Pickup tube

26. The foam concentrate inlet to an in-line foam eductor should not be more than ___ above the liquid surface of the foam concentrate. *(510)*

A. 3 feet *(1 m)*

B. 6 feet *(1.8 m)*

C. 9 feet *(2.8 m)*

D. 12 feet *(3.6 m)*

27. Which of the following is **not** a typical apparatus-mounted foam proportioning system? *(511)*

   A. Installed in-line eductor
   B. Around-the-pump proportioner
   C. Balanced pressure proportioner
   D. Apparatus recharging foam former

28. Firefighter A says that foam nozzle eductors are preferable to in-line foam eductors because they allow the attack team to easily relocate as fire conditions change.

   Firefighter B says that foam nozzle eductors are preferable to in-line foam eductors because they are built into the hoseline rather than into the nozzle.

   Who is right? *(510)*

   A. Firefighter A
   B. Firefighter B
   C. Both A and B
   D. Neither A nor B

29. The process of aerating foam is primarily achieved in what part of the system? *(511)*

   A. Foam proportioner
   B. Foam delivery device
   C. Fire hose
   D. Supply pump

30. A fog nozzle would be acceptable for foam application in which of the following situations? *(512)*

   A. AFFF and Class A foam application
   B. FFFP and Class B foam application
   C. Polar solvent fires
   D. Hydrocarbon fires and hazardous material spills

31. Which type of nozzle provides maximum expansion of the foam agent? *(512)*

   A. Solid bore nozzle
   B. Fixed-flow fog nozzle
   C. Automatic fog nozzle
   D. Air-aspirating foam nozzle

32. Firefighter A says that water-aspirating type nozzles use aerated water to produce a higher-air-volume foam than do mechanical blower generators.

   Firefighter B says that mechanical blower generators are used to produce high-expansion foam typically associated with total-flooding applications.

   Who is right? *(513)*

   A. Firefighter A
   B. Firefighter B
   C. Both A and B
   D. Neither A nor B

33. Which of the following is **not** a common reason for a failure to generate foam or for generating poor-quality foam in a system? *(513)*

   A. Eductor and nozzle flow ratings do not match, so foam concentrate cannot induct into the fire stream.
   B. Airtight fittings cause back pressure.
   C. Improper cleaning of proportioning equipment causes clogged foam passages.
   D. Hose lay on the discharge side of the eductor is too long, creating excess back pressure.

34. What method of foam application involves applying foam at the edges of a liquid pool? *(514)*

   A. Plunging   C. Bank-down

   B. Roll-on   D. Rain-down

35. What is the primary method of foam application for aboveground storage tank fires? *(514)*

   A. Bank-down   C. Plunging

   B. Rain-down   D. Roll-on

36. Firefighter A says that the bank-down method of foam application is particularly useful for dike fires and for fires involving overturned transport vehicles when an object is available above the liquid.

   Firefighter B says that with the roll-on method a firefighter continues to apply foam until it spreads over the entire surface of the fuel and the fire is extinguished.

   Who is right? *(514)*

   A. Firefighter A

   B. Firefighter B

   C. Both A and B

   D. Neither A nor B

37. What is the recommended approach for using the rain-down method of foam application on a large fire? *(514)*

   A. Apply foam at one edge and work toward the middle.

   B. Apply foam at one location and work out from that point.

   C. Apply foam to the far side of the fire and work back toward the attack team.

   D. Apply foam to the front edge of the fire and work back toward the fuel source.

38. What is the level of health risk posed to firefighters who handle foam concentrates? *(515)*

   A. There are no health risks from approved concentrates.

   B. Minimal risks due to skin and eye irritation; potentially harmful effects from inhalation or ingestion

   C. Serious risks that require the use of personal protective clothing at all times

   D. Extreme risks that require the use of personal protective clothing and SCBA at all times

39. What property of decomposition can lead to the death of marine creatures in waters contaminated by foams? *(515)*

   A. Reduced oxygen

   B. Increased oxygen

   C. Reduced temperature

   D. Increased temperature

40. Which of the following is **not** a correct statement with regard to foams and their environmental impact? *(515)*

   A. In the United States, Class A foams should be approved by the USDA Forest Service for environmental suitability.

   B. Foams that require larger amounts of oxygen to degrade have less environmental impact than those that require less oxygen.

   C. Generally, protein-based foams are safer for the environment.

   D. Manufacturers can provide information about the environmental impact of their products.

41. What term is applied to the action of a fire fighting foam in preventing the release of flammable vapors? *(498)*

   A. Separating
   B. Cooling
   C. Educting
   D. Suppressing

42. What is the mixture of foam concentrate with water prior to the introduction of air called? *(499)*

   A. Foam concentrate
   B. Nonaerated foam
   C. Finished foam
   D. Foam solution

43. In good quality foam, the bubbles should ___. *(499)*

   A. Be uniform in size
   B. All be very large
   C. All be very small
   D. Vary greatly in size

44. Firefighter A says that Class B foams designed for hydrocarbon fuels can be used effectively on polar solvent fuels if the concentration is increased.

   Firefighter B says that Class B foams designed for polar solvent fuels can be used effectively on hydrocarbon fuels if the manufacturer of the foam specifies such use.

   Who is right? *(500)*

   A. Firefighter A
   B. Firefighter B
   C. Both A and B
   D. Neither A nor B

45. Which of the following statements is true with regard to Class A foams? *(500)*

   A. Class A foams have supercleaning characteristics making it unnecessary to thoroughly flush equipment after use.

   B. Class A foams are mildly corrosive.

   C. An advantage of Class A foams is that they reduce water penetration into fuels so that burning liquids do not float to the top of attack sprays.

   D. Class A foams cannot be used with medium- and high-expansion devices.

46. Proper proportioning of a foam concentrate to achieve a 6% concentration requires ___. *(505)*

   A. 6 parts of water for each 94 parts of foam concentrate

   B. 6 parts of foam concentrate for each 94 parts of water

   C. 6 parts of water for each 100 parts of foam concentrate

   D. 6 parts of foam concentrate for each 100 parts of water

47. How should a firefighter produce a thick foam for exposure protection with a Class A foam concentrate? *(505)*

    A. Close the nozzle opening.

    B. Widen the nozzle opening.

    C. Reduce the amount of water used.

    D. Reduce the amount of concentrate used.

48. Which of the following statements is true with regard to Class B foam proportioning? *(507)*

    A. All Class B foams are designed to be used on either hydrocarbon or polar solvent fuels, but not both.

    B. Medium-expansion Class B foams are normally mixed at concentrations of 6% or higher.

    C. Some multipurpose Class B foams are designed for use on polar solvent fuels at 3% concentrations or for use on hydrocarbon fuels at 6% concentrations.

    D. Some multipurpose Class B foams are designed to be applied at 3% concentrations on either hydrocarbon fuels or polar solvent fuels.

49. The only use of the roll-on method of foam application is on ___. *(514)*

    A. A pool of ignited liquid fuel on the open ground

    B. A pool of unignited liquid fuel on the open ground

    C. A pool of liquid fuel (either ignited or unignited) on the open ground

    D. A pool of liquid fuel (either ignited or unignited) that has been diked

50. Firefighter A says that the rate of Class B foam application is not affected by whether or not the fuel has ignited.

    Firefighter B says that the rate of Class B foam application is affected by whether the fuel has spilled or is in a tank, as well as what type of tank is involved.

    Who is right? *(504)*

    A. Firefighter A

    B. Firefighter B

    C. Both A and B

    D. Neither A nor B

REVIEW TEST ANSWER SHEET

	A	B	C	D
1.	○	○	○	○
2.	○	○	○	○
3.	○	○	○	○
4.	○	○	○	○
5.	○	○	○	○
6.	○	○	○	○
7.	○	○	○	○
8.	○	○	○	○
9.	○	○	○	○
10.	○	○	○	○
11.	○	○	○	○
12.	○	○	○	○
13.	○	○	○	○
14.	○	○	○	○
15.	○	○	○	○
16.	○	○	○	○
17.	○	○	○	○
18.	○	○	○	○
19.	○	○	○	○
20.	○	○	○	○
21.	○	○	○	○
22.	○	○	○	○
23.	○	○	○	○
24.	○	○	○	○
25.	○	○	○	○
26.	○	○	○	○
27.	○	○	○	○
28.	○	○	○	○
29.	○	○	○	○
30.	○	○	○	○
31.	○	○	○	○
32.	○	○	○	○
33.	○	○	○	○

	A	B	C	D
34.	○	○	○	○
35.	○	○	○	○
36.	○	○	○	○
37.	○	○	○	○
38.	○	○	○	○
39.	○	○	○	○
40.	○	○	○	○
41.	○	○	○	○
42.	○	○	○	○
43.	○	○	○	○
44.	○	○	○	○
45.	○	○	○	○
46.	○	○	○	○
47.	○	○	○	○
48.	○	○	○	○
49.	○	○	○	○
50.	○	○	○	○
51.	○	○	○	○
52.	○	○	○	○
53.	○	○	○	○
54.	○	○	○	○
55.	○	○	○	○
56.	○	○	○	○
57.	○	○	○	○
58.	○	○	○	○
59.	○	○	○	○
60.	○	○	○	○
61.	○	○	○	○
62.	○	○	○	○
63.	○	○	○	○
64.	○	○	○	○
65.	○	○	○	○
66.	○	○	○	○
67.	○	○	○	○

	A	B	C	D
68.	○	○	○	○
69.	○	○	○	○
70.	○	○	○	○
71.	○	○	○	○
72.	○	○	○	○
73.	○	○	○	○
74.	○	○	○	○
75.	○	○	○	○
76.	○	○	○	○
77.	○	○	○	○
78.	○	○	○	○
79.	○	○	○	○
80.	○	○	○	○
81.	○	○	○	○
82.	○	○	○	○
83.	○	○	○	○
84.	○	○	○	○
85.	○	○	○	○
86.	○	○	○	○
87.	○	○	○	○
88.	○	○	○	○
89.	○	○	○	○
90.	○	○	○	○
91.	○	○	○	○
92.	○	○	○	○
93.	○	○	○	○
94.	○	○	○	○
95.	○	○	○	○
96.	○	○	○	○
97.	○	○	○	○
98.	○	○	○	○
99.	○	○	○	○
100.	○	○	○	○

Name _____

Date _____

Score _____

Chapter 13 Competency Profile

Student Name _____ Soc. Sec. No. _____
Last   First   Middle

Fire Department _____

Address _____

Phone _____

Home Address _____

Phone _____

Date of Enrollment _____ - _____ - _____   Total Class Hours _____

Date of Withdrawal _____ - _____ - _____   Total Hours Absent _____

Date of Completion _____ - _____ - _____

Instructor's Name _____   Session Dates _____

Instructor's Directions

1. Check the candidate's competency rating (3, 2, 1, ☒) for each performance test task and psychomotor lesson objective (practical activity and job sheets) listed below.

2. List any additional performance tasks or psychomotor objectives (job sheets or practical activity sheets) under "Other," and check the candidate's competency rating.

3. Record the candidate's cognitive scores (written lesson tests and *administered* chapter review tests) in the spaces provided.

Level				Psychomotor Competencies
3	2	1	☒	

Practical Activity Sheets

3	2	1	☒	
☐	☐	☐	☐	PAS 13-1 — Select Foams for Specific Fire Situations
☐	☐	☐	☐	PAS 13-2 — Select Nozzles for Specific Fire Situations
☐	☐	☐	☐	Other _____
☐	☐	☐	☐	_____

Job Sheets

3	2	1	☒	
☐	☐	☐	☐	JS 13-1 — Install an In-Line Eductor and Operate a High-Expansion Foam Generator
☐	☐	☐	☐	Other _____
☐	☐	☐	☐	_____

Chapter 13 Performance Test

3	2	1	☒	
☐	☐	☐	☐	Task 1 — Extinguish a Class B fire with foam.

Level			
3	2	1	☒
☐	☐	☐	☐
☐	☐	☐	☐
☐	☐	☐	☐

Psychomotor Competencies

Task 2 — Extinguish a fire involving a fuel spill from a passenger vehicle.

Other _____

_____

Points Achieved	Points Needed/ Total	**Cognitive Competencies**

Written Test

_____	3/3	1. Describe the basic methods by which foam prevents or controls a hazard.
_____	8/10	2. Classify flammable liquids as hydrocarbon or polar solvent fuels.
_____	8/8	3. Explain how foam is generated.
_____	4/4	4. Describe the components of foam production.
_____	3/4	5. List factors that affect foam expansion.
_____	2/3	6. Classify foams by their expansion ratios.
_____	10/12	7. Distinguish between characteristics of Class A and Class B foams.
_____	4/5	8. List factors that affect Class B foam application rates.
_____	4/5	9. Select facts about proportioning.
_____	4/4	10. Match methods of proportioning to their descriptions.
_____	4/5	11. Select facts about proportioners.
		12. Evaluated on Practical Activity Sheet 13-1
_____	4/5	13. Match types of handline foam nozzles to their uses.
		14. Evaluated on Practical Activity Sheet 13-2
_____	4/5	15. List reasons for poor foam generation.
_____	3/3	16. Match foam application methods to their uses.
_____	3/3	17. List types of hazards associated with foam use.
		18. Evaluated on Job Sheet 13-1

Review Test

Chapter 13 Review Test

_____

Instructor's Signature _____ Date _____

Student's Signature _____ Date _____

STUDENT APPLICATIONS

FOURTH EDITION

ESSENTIALS OF FIRE FIGHTING

LESSON
14

IGNITABLE LIQUID &
FLAMMABLE GAS
FIRE CONTROL

FIREFIGHTER II

FIRE PROTECTION PUBLICATIONS
OKLAHOMA STATE UNIVERSITY

fpp

Study Objectives

LESSON OBJECTIVE

After completing this lesson, you will be able to operate as part of a team to coordinate an interior attack and to control and/or extinguish ignitable liquid fires and flammable gas cylinder fires.

ENABLING OBJECTIVES

After reading Chapter 14 of **Essentials,** pages 529 through 535, and completing related activities, you will be able to —

1. Distinguish between flammable liquids and combustible liquids.

2. Select facts about suppressing Class B fires.

3. Describe signs and effects of BLEVE.

4. List the four ways that water can be used to attack a Class B fire.

5. List methods of identifying tank contents.

6. Select facts about techniques for suppressing bulk transport vehicle fires.

7. **Use water to control an ignitable liquid fire in an open pan. *(Job Sheet 14-1)***

8. Distinguish between the characteristics of natural gas and liquid petroleum gas.

9. **Control and/or extinguish a flammable gas cylinder fire. *(Job Sheet 14-2)***

10. **Determine actions to take, including retreat, when dealing with specific Class B fire conditions. *(Practical Activity Sheet 14-1)***

Study Sheet

Introduction

This study sheet is intended to help you learn the material in Chapter 14 of *Essentials of Fire Fighting,* Fourth Edition. You may use it for self-study, or you may use it to review material that will be covered in the lesson and chapter review tests. The numbers in parentheses are the pages in *Essentials* on which the answers or terms can be found.

Chapter Vocabulary

Be sure you know the chapter-related meanings of the following terms and abbreviations. Use a dictionary or the glossary in *Fire Service Orientation and Terminology* if you cannot determine the meaning of the term from its context:

- Asphyxiant *(533)*
- BLEVE *(530)*
- Combustible liquid *(529)*
- Flammable liquid *(529)*
- Hydrocarbon liquids *vs.* polar solvents *(529)*
- LPG *vs.* natural gas *(533, 534)*
- Manifest *(533)*
- Mercaptan *(534)*
- Placard *(533)*
- Wicking *(530)*

Study Questions & Activities

1. Distinguish between a flammable liquid and a combustible liquid. *(529)*

   a. Flammable liquid _____

   _____

   b. Combustible liquid _____

   _____

2. Distinguish between a hydrocarbon and a polar solvent. *(529)*

   a. Hydrocarbon _____

   _____

   b. Polar solvent _____

   _____

3. Why is it important for firefighters to be cautious about standing in pools of fuel or water runoff containing fuel? *(530)*

   _____

   _____

4. What needs to be done before fires burning around relief valves or piping can be extinguished? *(530)*

_____

_____

_____

5. What assumption should firefighters make about protection from relief valves? *(530)*

_____

_____

_____

_____

6. When does a BLEVE most commonly occur? How should these fires be attacked? *(530)*

_____

_____

_____

_____

7. What is the preferred method of controlling flammable liquid fires? *(530)*

_____

_____

8. Describe the following ways in which water can be used in a Class B fire. *(531, 532)*
   a. Cooling agent _____

   _____

   _____

   b. Mechanical tool _____

   _____

   _____

   c. Substitute medium _____

   _____

   _____

   d. Protective cover _____

   _____

   _____

9. List the additional difficulties involved in extinguishing fires in vehicles transporting flammable fuels as opposed to those in storage facilities. *(532)*

   _____

   _____

   _____

   _____

10. What are some procedures that should be followed when handling vehicle fires? *(532, 533)*

_____

_____

_____

_____

11. Name some ways that firefighters can determine the exact nature of a transport vehicle's cargo. *(533)*

_____

_____

_____

12. Why is it important for firefighters to have a working knowledge of the hazards and correct procedures in handling incidents related to natural gas and LPG? *(533)*

_____

_____

_____

13. Discuss the differences between natural gas and LPG. *(533, 534)*

_____

_____

_____

_____

_____

14. What is the responsibility of the local utility company when an emergency involving natural gas occurs? What is the responsibility of firefighters at the scene of such an incident? *(533)*

a. Utility_____

_____

_____

b. Firefighters_____

_____

_____

_____

15. If gas is burning as the result of a broken main, why should the flame not be extinguished? What should be done instead? *(534, 535)*

_____

_____

_____

_____

Practical Activity Sheet 14-1
Determine Actions to Take, Including Retreat, When Dealing with Specific Class B Fire Conditions

Name _____ Date _____

Evaluator _____ Overall Competency Rating _____

References	NFPA 1001, Fire Ground Operations 4-3.3 *Essentials*, pages 529–535
Prerequisites	None
Introduction	Class B fires present many situations that firefighters do not commonly encounter, such as volatile fuels, containment problems, the dangers of explosive vapors and BLEVE, etc. For these reasons, firefighters must be prepared through training to deal with such incidents. This assignment will give you the opportunity to practice Class B fire tactics.
Directions	Study the following scenarios and answer the questions that accompany each.
Activity	SCENARIO 1 A grass fire has spread into a bulk oil storage area. Though there is little fuel in the storage area, the heat is intense near the fence around the storage area. 1. What should the attack team do with regard to the bulk storage area? _____ _____ 2. If water is applied to the storage tanks, where should the water be directed? _____ _____ 3. A leak develops in the outlet pipe of one tank. The leak is small but represents a danger because it is near the point of fire. How can the attack team use water to control this leak? _____ _____ _____

4. Plant personnel advise the attack team that there is a shutoff valve between the tank and the outlet pipe leak. Command determines that the valve should be closed to avoid exposing crude oil to the fire. How should the response team be protected while turning off the valve?

_____

_____

_____

_____

5. At one point in the response, a relief valve opens and a bulge appears near the roof seam of the storage tank. What might these conditions indicate?

_____

_____

_____

SCENARIO 2

A gasoline transport truck has overturned on a busy, divided four-lane street. Fuel is leaking but not ignited. The gasoline is flowing downhill toward a storm sewer drain. The vehicle is on its side in the median between the two sections of road. The driver has escaped the accident unharmed.

1. What action should the response team take in controlling traffic?

_____

_____

2. If two engines have responded, what considerations should be made in placement of the apparatus?

_____

_____

3. To the extent possible, where should the team position themselves relative to topography, weather conditions, and street?

   a. _____

   b. _____

   c. _____

4. Should the team try to direct the flow of gasoline into the storm drain? Why or why not?

_____

_____

_____

_____

5. Part of the fuel is spreading across the road, making it impossible to allow traffic to pass without driving into the fuel. Command determines that the fuel should be contained with a fuel stream so that the flow of traffic can continue. How can water be used to control the spill?

_____

_____

_____

Scenario 3

An excavation crew installing television cable lines has ruptured an underground natural gas pipeline. The rupture immediately ignited, engulfing the backhoe and seriously burning the equipment operator.

1. What should the team do immediately with regard to the area around the incident?

    a. _____

    b. _____

    c. _____

2. From what direction should the team approach the incident?

    _____

    _____

3. Should the team try to extinguish the fire at the leak?

    _____

    _____

4. One of the team members notices that a few yards *(meters)* away, there is a fenced area with an exposed valve. A sign on the fence states: "Danger: Natural Gas Control Valve. Keep Away. No Smoking." What should the response team do with regard to the valve?

    _____

    _____

    _____

5. What outside resource should the department contact?

    _____

    _____

    _____

Competency Rating Scale

3 — Skilled — All 15 questions answered appropriately per responses suggested in instructor's answer guide; student requires no additional practice.

2 — Moderately skilled — Correctly answered at least 4 questions for each scenario; student may benefit from additional practice.

1 — Unskilled — Correctly answered fewer than 4 questions for each scenario; student requires additional practice and reevaluation.

☒ **—Unassigned** — Task is not required or has not been performed.

✔ **Evaluator's Note:** Evaluate the results or product as indicated below. Calculate the percentage of correct responses, and assign an overall competency level. Record this competency rating on both the student's practical activity sheet and competency profile.

To show competency in this objective, the student must achieve an overall rating of at least 2.

Answers	Correct	Incorrect
All answers evaluated per answers in **Instructor's Guide.**		
SCENARIO 1		
Question 1	☐	☐
Question 2	☐	☐
Question 3	☐	☐
Question 4	☐	☐
Question 5	☐	☐
SCENARIO 2		
Question 1	☐	☐
Question 2	☐	☐
Question 3	☐	☐
Question 4	☐	☐
Question 5	☐	☐
SCENARIO 3		
Question 1	☐	☐
Question 2	☐	☐
Question 3	☐	☐
Question 4	☐	☐
Question 5	☐	☐

Job Sheet 14-1
Use Water to Control an Ignitable Liquid Fire in an Open Pan

Name _____ Date _____

Evaluator _____ Overall Competency Rating _____

References	NFPA 1001, Fire Ground Operations 4-3.3 ***Essentials,*** pages 529–535
Prerequisites	None
Candidate's Instructions	To meet evaluation standards, you must perform this job within _____ *[amount of time, if applicable]*; you may have _____ attempts. When you are ready to perform this job, ask your instructor to observe the procedure and complete this form. To show mastery of this job, you must perform all steps to receive an overall competency rating of at least 2.

Competency Rating Scale

3 — Skilled — Meets all evaluation criteria and standards; performs task independently on first attempt; requires no additional practice or training.

2 — Moderately skilled — Meets all evaluation criteria and standards; performs task independently; additional practice is recommended.

1 — Unskilled — Is unable to perform the task; additional training required.

☒ — **Unassigned** — Job sheet task is not required or has not been performed.

✔ **Evaluator's Note:** Formulate and inform the candidate of the standards for this task (time allowed and number of attempts). Observe the candidate perform the task, check the step/key point under the appropriate attempt number as accomplished, record total time (if appropriate), and then use the rating scale above to assign an overall competency rating. If the candidate is unable to perform any step of this job, have the candidate review the materials and try again.

Introduction	Every firefighter is familiar with the phrase, "Never put water on flammable liquid fires," yet necessity and experience have shown that water is highly effective in controlling or extinguishing these fires. For instance, water can be used as a cooling agent, and it can be used as a mechanical tool to move the fuel to areas where it can safely burn or where ignition sources are more easily controlled. Water may also be used as a substitute medium to displace fuel from pipes or tanks that are leaking, and as a protective cover for teams advancing to shut off liquid or gaseous fuels. In this job sheet you will use water as a cooling agent and as a mechanical tool to move the burning fuel to an area where it can safely burn or where ignition sources are more easily controlled.

	Equipment and Personnel
Equipment and Personnel	• At least six firefighters, three on each attack line, dressed in full protective clothing and SCBA • Two 1½-inch *(38 mm)* or larger charged attack lines equipped with fog nozzles • One 1½-inch *(38 mm)* or larger charged backup line supplied from a second water source • Class B fire set up in a pan and monitored according to *NFPA 1406*

✔**Note to Instructor:** Ensure firefighter safety at all times during this training evolution. Before proceeding with live fire training evolutions, read and adhere to *NFPA 1406 Standard on Outside Live Fire Training Evolutions.* A backup line of the same or larger size should always be manned and charged when conducting live fire training. It should be supplied from a second water source or pump.

Have students repeat this exercise, rotating the hoseline duties so that each student has a chance to perform on the nozzle.

Job Steps	Key Points	Attempt No. 1	2	3
1. Position the hoselines.	1. a. Parallel to each other	—	—	—
	b. About 3 to 5 feet *(1 m to 1.5 m)* apart	—	—	—
	c. Out of fire area	—	—	—
	d. Nozzles toward fire	—	—	—
2. Position yourselves.	2. a. At least three firefighters to each attack line	—	—	—
	b. Nozzle firefighters in foremost position	—	—	—
	c. Backup firefighters an arm's length from nozzle firefighters	—	—	—
	d. Remaining firefighters about 3 feet *(1 m)* apart	—	—	—
	e. On inside of hose	—	—	—
(Firefighters on nozzles) 3. Set the nozzle flow.	3. a. For a wide-angle or penetrating fog pattern	—	—	—
	b. For gallonage (if applicable) indicated by your instructor	—	—	—
4. Open the nozzle.	4. a. Fully	—	—	—
	b. Briefly	—	—	—
	c. Aiming stream to side	—	—	—
	d. Testing fog pattern	—	—	—
	e. Expelling air	—	—	—

Job Steps	Key Points	Attempt No.		
		1	2	3
5. Start your SCBA airflow.	5. All members of team **!CAUTION:** A safety officer will check your gear before you enter the fire area. Make sure that *no skin or hair is exposed!*	___	___	___
(Instructor) 6. Position yourself.	6. a. Between teams	___	___	___
	b. Hand on near shoulder of each nozzle firefighter	___	___	___
7. Give the command to open the nozzles.	7. When the teams are in place	___	___	___
8. Give the command to attack the fire.	8. When fog streams from both teams overlap slightly at center providing a protective barrier	___	___	___
9. *(Both teams)* Advance the line.	9. a. On instructor's command	___	___	___
	b. Sliding your feet in unison	___	___	___
	c. Keeping fog streams overlapped slightly in center	___	___	___
	d. Until within stream reach of fire	___	___	___
(Nozzle firefighters) 10. Narrow and widen the fog streams.	10. As necessary to control the fire	___	___	___
11. Sweep the fire out.	11. a. Slowly	___	___	___
	b. From side to side	___	___	___
	c. Leading edge of fog patterns in contact with fuel surface at all times	___	___	___
12. Reset stream pattern if necessary	12. a. To wide-angle fog pattern	___	___	___
	b. Streams overlapping slightly at center	___	___	___
13. Exit the fire area.	13. a. Backing up with a shuffle step	___	___	___
	b. Taking up slack hose to prevent falling or tripping	___	___	___

Job Steps	Key Points	Attempt No.		
		1	2	3
14. Shut the nozzles.	14. When all personnel are out of fire area	___	___	___
15. Stop SCBA airflow and doff SCBA masks.	15. When well clear of fire area	___	___	___
	Time (Total)	___	___	___

Evaluator's Comments _____

_____

_____

Job Sheet 14-2
Control and/or Extinguish a Flammable Gas Cylinder Fire

Name _____ Date _____

Evaluator _____ Overall Competency Rating _____

References	NFPA 1001, Fire Ground Operations 4-3.3 *Essentials,* pages 529–535
Prerequisites	None
Candidate's Instructions	To meet evaluation standards, you must perform this job within _____ *[amount of time, if applicable]*; you may have _____ attempts. When you are ready to perform this job, ask your instructor to observe the procedure and complete this form. To show mastery of this job, you must perform all steps to receive an overall competency rating of at least 2.

Competency Rating Scale

3 — Skilled — Meets all evaluation criteria and standards; performs task independently on first attempt; requires no additional practice or training.

2 — Moderately skilled — Meets all evaluation criteria and standards; performs task independently; additional practice is recommended.

1 — Unskilled — Is unable to perform the task; additional training required.

☒ **—Unassigned** — Job sheet task is not required or has not been performed.

✔ **Evaluator's Note:** Formulate and inform the candidate of the standards for this task (time allowed and number of attempts). Observe the candidate perform the task, check the step/key point under the appropriate attempt number as accomplished, record total time (if appropriate), and then use the rating scale above to assign an overall competency rating. If the candidate is unable to perform any step of this job, have the candidate review the materials and try again.

Introduction	Response to a flammable gas cylinder fire requires a careful and well coordinated attack. As a minimum, water will be employed as an extinguishing agent and as a cooling agent. Water streams may also be used to provide protective cover for teams assigned to shut off the fuel. There are three ways that flammable gas cylinders can be involved in a fire. The container may be exposed to flames from fire fed by another fuel, such as gasoline around an overturned bulk tank transport or a grass fire near LPG storage tanks. Flammable gas cylinder fires also include those for which the container contents are the fuel. Finally, a flammable gas cylinder fire may involve multiple fuel sources, including cylinder contents and fuels exterior to the cylinder.

The primary tactical objective in making an attack on a fire involving a flammable gas cylinder is to keep the container and its supports cool while the fire is being attacked. This not only protects exposures, but it also reduces the possibility of BLEVE and structural rupture due to failure or collapse of the tank.

If the fire involves the contents of the tank, such as flames coming from a ruptured pipeline, a second objective is to stop the leak. If possible, a shutoff valve will be used to stop the leak, though water may also be used as a substitute medium to help contain a leak. Firefighters must exercise caution to avoid extinguishing the fire until the leak is eliminated as the fuel is generally less dangerous burning than it would be if the fuel were allowed to spread with the risk of having the fuel encounter an ignition source. Teams advancing to shut off leaks must be well protected.

Equipment and Personnel

- Nine firefighters, three on the attack line and three on each protective cover line, dressed in full protective clothing and SCBA
- Two or more charged master stream devices equipped with solid stream nozzles
- Three 1½-inch *(38 mm)* or larger charged attack lines equipped with fog nozzles
- One 1½-inch *(38 mm)* or larger charged backup line supplied from a second water source
- Flammable gas cylinder fire set up and monitored according to *NFPA 1406*

✔**Note to Instructor:** Ensure firefighter safety at all times during this training evolution. Before proceeding with live fire training evolutions, read and adhere to *NFPA 1406 Standard on Outside Live Fire Training Evolutions.* A backup line of the same or larger size should always be manned and charged when conducting live fire training. It should be supplied from a second water source or pump.

Have students repeat this exercise, rotating the hoseline duties so that each student has a chance to perform on the nozzle.

Job Steps	Key Points	Attempt No. 1	2	3
1. *(Two firefighters)* Deploy the master stream device.	1. a. At the minimum safe distance from cylinder	—	—	—
	b. Placed to provide full coverage of top of cylinder	—	—	—
	c. Securely anchored	—	—	—
	d. Backing away after anchoring device	—	—	—
2. Supply water to the master stream device.	2. a. Desired stream and gallonage (if applicable) as indicated by your instructor	—	—	—
	b. Decreasing pressure at supply source if device starts to move	—	—	—

(Attack team)

3. Position the two attack hoselines.

   3. a. Parallel to each other   ___ ___ ___

        b. About 3 to 5 feet *(1 m to 1.5 m)* apart   ___ ___ ___

        c. Out of fire area   ___ ___ ___

        d. Nozzles toward fire   ___ ___ ___

4. Position yourselves.

   4. a. At least three firefighters to each line   ___ ___ ___

        b. Backup firefighters an arm's length from nozzle firefighters   ___ ___ ___

        c. Remaining firefighters about 3 feet *(1 m)* apart   ___ ___ ___

        d. On inside of both hoselines   ___ ___ ___

(Firefighters on attack nozzles)

5. Adjust the nozzle.

   5. a. Wide fog pattern   ___ ___ ___

        b. Gallonage (if applicable) indicated by your instructor   ___ ___ ___

(Firefighter on backup nozzle)

6. Adjust the nozzle.

   6. a. Fog pattern   ___ ___ ___

        b. Gallonage (if applicable) indicated by your instructor   ___ ___ ___

(All three nozzle firefighters)

7. Open the nozzle.

   7. a. Fully   ___ ___ ___

        b. Briefly   ___ ___ ___

        c. Aiming stream to side   ___ ___ ___

        d. Testing fog pattern   ___ ___ ___

        e. Expelling air   ___ ___ ___

(Attack team and backup)

8. Start your SCBA airflow.

   8. All members of team   ___ ___ ___

        !CAUTION: A safety officer will check your gear before you enter the fire area. Make sure that ***no skin or hair is exposed!***

(Instructor)

9. Position yourself.

   9. a. Between attack teams   ___ ___ ___

        b. Hand on near shoulder of each nozzle firefighter   ___ ___ ___

10. Give the command to open the nozzles.

   10. When the teams are in place   ___ ___ ___

Job Steps	Key Points	Attempt No. 1 2 3
11. Give the command to attack the fire.	11. a. When fog streams from both attack lines overlap slightly at center	— — —
	b. When bottom edges of fog streams touch ground	— — —
(Attack team) 12. Advance the line.	12. a. On instructor's command	— — —
	b. Approaching cylinder from the side	— — —
	c. Sliding your feet in unison in slow, deliberate movement	— — —
	d. Keeping fog streams overlapped and bottom edges touching ground	— — —
	e. Remaining behind protective barrier	— — —
	f. Until within area of shut off valve area	— — —
	g. Progressively widening fog pattern	— — —
13. Close the cylinder's shutoff valve.	13. (None)	— — —
14. Exit the fire area.	14. a. When the shutoff valve has been closed, the leak has been stopped, if signs of BLEVE develop, or if the fire is extinguished prematurely	— — —
	b. Backing up with a shuffle step	— — —
	c. Taking up slack hose to prevent falling or tripping	— — —
15. *(Three nozzle firefighters)* Shut the nozzles.	15. When all personnel are out of fire area	— — —
16. Stop SCBA airflow and doff SCBA masks.	16. When well clear of fire area	— — —

Time (Total) — — —

Evaluator's Comments _____

_____

_____

Chapter 14 Review Test

> ➡ **Directions:** This review test covers the Firefighter II material in Chapter 14 of your ***Essentials of Fire Fighting*** text. It may be assigned as a study aid (self-test) or may be administered by your instructor as a pretest or posttest.
>
> When used as a study aid, try to answer the questions without referring to the page numbers in ***Essentials*** or your ***Firefighter II Student Applications*** workbook *(SA)* on which the answers can be found until after you have completed the entire test. Then check your answers against those on the pages provided in parentheses.
>
> When administered by your instructor as a pretest or posttest, read each of the test questions carefully. Choose the best response and then darken the corresponding letter on your answer sheet.
>
> This chapter review test contains 33 multiple-choice questions, each worth 3 points. To pass the test, you must achieve at least 84 of the 99 points possible.

1. What class of fires is defined as those involving flammable and combustible liquids and gases? *(529)*

   A. Class A   C. Class C

   B. Class B   D. Class D

2. Substances that are normally in a liquid state and that have a flash point less than 100°F *(38°C)*, such as gasoline and acetone, are known as ___. *(529)*

   A. Ignitable liquids

   B. Combustible liquids

   C. Noncombustible liquids

   D. Flammable liquids

3. Substances that are normally in a liquid state and that have a flash point greater than 100°F *(38°C)*, such as kerosene and vegetable oil, are known as ___. *(529)*

   A. Ignitable liquids

   B. Combustible liquids

   C. Inflammable liquids

   D. Flammable liquids

4. Firefighter A says that liquid fuels that mix with water are known as hydrocarbons.

   Firefighter B says that liquid fuels that do not mix with water are known as polar solvents.

   Who is right? *(529)*

   A. Firefighter A   C. Both A and B

   B. Firefighter B   D. Neither A nor B

5. Which of the following is **not** a danger presented by firefighters standing in pools of liquid fuels? *(530)*

   A. During cold weather, the fuel may freeze and become slick.

   B. The fuel can ignite and engulf the firefighters.

   C. Fuel can wick up into protective clothing, making it flammable.

   D. Fuel can wick up into protective clothing and cause contact burns on the skin.

6. What action should be taken for flowing-fuel fires burning around relief valves or piping if the leaking product cannot be turned off? *(530)*

   A. Extinguish the fire as a first priority.

   B. Dispatch personnel in full thermal gear to repair the leak.

   C. Set up a master foam nozzle to keep the site of the leak in a constant flow of foam.

   D. Do not extinguish the fire but try to contain the pooling fuel.

7. Which of the following is an accurate statement with regard to fires involving liquid fuels? *(530)*

   A. Unburned vapors are generally lighter than air and will quickly disperse in the open or collect at the ceiling in enclosed areas.

   B. An increase in the intensity of the sound or fire issuing from a relief valve indicates that the container is nearly empty and that most of the fuel has escaped.

   C. Relief valves may not be adequate to relieve excess pressures from systems exposed to severe fire conditions.

   D. If properly grounded, vehicles and electrical systems do not present ignition sources in the presence of flammable vapors.

8. What does the abbreviation *BLEVE* stand for? *(530)*

   A. Boiling liquid expanding vapor explosion

   B. Balanced level exhaust valve eruption

   C. Class B liquid explosive vapor event

   D. Burning liquid emergency victim evacuation

9. What is the recommended method of fighting a fire when flames contact the flammable liquid vessel above the liquid level? *(530)*

   A. Apply foam to engulf the container.

   B. Apply water to the lower portion of the container to cool the liquid.

   C. Apply water to the upper portion of the container.

   D. Apply water to the base of the fire as far as practical from the container.

10. Firefighter A says that foam is the preferred method for controlling flammable liquid fires.

    Firefighter B says that water can be used for flammable liquid fires as a cooling agent, mechanical tool, substitute medium, and to provide protective cover.

    Who is right? *(530)*

    A. Firefighter A   C. Both A and B

    B. Firefighter B   D. Neither A nor B

11. Which of the following statements is true with regard to using water as a cooling agent for flammable liquid fires? *(531)*

   A. Water should not be used as a cooling agent for flammable liquid fires.

   B. Water is most effective on heavier oils.

   C. Water is most effective on light petroleum distillates.

   D. Because of the intense heat of flammable liquid fires, water is ineffective as a cooling agent for exposures.

12. What is the recommended method for using a water stream to move liquid fuels (burning or not) to areas where they can be more easily extinguished or controlled? *(531)*

   A. Water streams should not be used to move liquid fuels.

   B. Plunge the stream into the leading edge of the spill.

   C. Sweep the stream back and forth with the trailing edge of the fog pattern in contact with the fuel surface.

   D. Sweep the stream back and forth with the leading edge of the fog pattern in contact with the fuel surface.

13. Which of the following is an acceptable method of using water to combat a small liquid fuel leak? *(531)*

   A. Apply water to the point where the leak contacts the ground to keep the fuel from pooling and to keep it diluted.

   B. Apply water directly to the leak site at a pressure greater than the leak.

   C. Apply water to a relief valve at a pressure greater than the relief pressure to reduce the likelihood of additional leaks.

   D. None of the above are acceptable; do not use water to combat liquid fuel leaks.

14. Which of the following is **not** an acceptable method of using water as a substitute medium for liquid fuel fires? *(531)*

   A. For large fires or spills, use water to dilute the fuel to a point that it is no longer flammable.

   B. Fill leaking containers with water so that the lighter fuel is raised above the leak.

   C. Pump water into leaking pipes to reverse the flow of fuel and hold it beyond the point of the leak.

   D. None of the above are acceptable; water should not be used as a substitute medium for liquid fuel fires.

15. What is the minimum preferred deployment of hoselines when providing protective cover to attack teams at a flammable liquid fire? *(532)*

   A. One hoseline and one backup line

   B. Two hoselines and one backup line

   C. Two unattended master stream devices

   D. None of the above is correct; water streams should not be used to provide protective cover at flammable liquid fires

16. When tanks of flammable liquids are exposed to flame impingement, what method should be used to control the situation? *(532)*

    A. Apply water to the base of the tank, sweeping back and forth slowly.

    B. Lob water to the top of the tank so that water runs down the sides.

    C. Apply water to the tank outlet and work the stream slowly to the inlet and then back to the outlet.

    D. None of the above is correct; water streams should not be used on tanks containing flammable liquids.

17. Firefighter A says that attack teams should approach flammable liquid storage vessels from the ends.

    Firefighter B says that attack teams should approach flammable liquid storage vessels at right angles to the sides of the vessels.

    Who is right? *(532)*

    A. Firefighter A   C. Both A and B

    B. Firefighter B   D. Neither A nor B

18. What is the recommended traffic pattern for an incident involving a flammable liquid transport vehicle? *(532)*

    A. Do not disrupt the flow of traffic.

    B. Close down the lanes involved in the incident.

    C. Close down the lanes involved in the incident and at least one additional lane.

    D. Shut down the flow of traffic from both directions completely.

19. Firefighter A says that an incident site involving a flammable liquid transport vehicle at night should be marked off with open-flame flares.

    Firefighter B says that if law enforcement personnel are not available at an incident site involving a flammable liquid transport vehicle, then a firefighter should be assigned to provide traffic control.

    Who is right? *(532, 533)*

    A. Firefighter A   C. Both A and B

    B. Firefighter B   D. Neither A nor B

20. What may an increase in the intensity of sound or fire issuing from a relief valve indicate? *(530)*

    A. The relief valve has opened.

    B. The vessel may be about to rupture.

    C. The cooling operation has been successful.

    D. The vessel is designed for BLEVE protection.

21. Which of the following statements about natural gas is true? *(533)*

    A. Natural gas is toxic and also displaces normal breathing air.

    B. Natural gas is lighter than air so it tends to rise and diffuse in the open.

    C. Natural gas is explosive in concentrations between 30 to 50 percent.

    D. Natural gas is piped at pressures of 1,000 to 7,000 psi *(7 000 kPa to 49 000 kPa)*.

22. For incidents involving natural gas, responding crews should seek aid from ___. (533)

   A. The Federal Emergency Management Agency to evacuate everyone within a one-mile (1.6 km) radius

   B. Local health agencies to handle victims of toxic gas exposure

   C. The local gas utility for a response crew with special equipment and knowledge of the gas distribution system

   D. The Environmental Protection Agency to assess potential environmental hazards

23. Which of the following statements is true of liquefied petroleum gas (LPG)? (534)

   A. LPG is composed mostly of methane.

   B. LPG is lighter than air than air.

   C. LPG is toxic.

   D. LPG is explosive in concentrations of 1.5 and 10 percent.

24. Unburned LPG should be dissipated with a fog stream of at least ___ . (534)

   A. 25 gpm (100 L/min)

   B. 50 gpm (200 L/min)

   C. 75 gpm (300 L/min)

   D. 100 gpm (400 L/min)

25. Which of the following is a guideline for dealing with incidents that involve gas lines? (534)

   A. If a gas line breaks, firefighters should check buildings in the area for gas buildup.

   B. If gas from a broken pipe is not burning, firefighters should immediately close main valves in the line.

   C. If a gas from a broken pipe is burning, the fire should be extinguished.

   D. Apparatus should approach the break at a right angle to the path of the pipe line.

26. What does the illustration below show? (535)

   A. A gas line petcock in the closed position

   B. A gas line petcock in the open position

   C. A gas line directional flow indicator

   D. A gas line rate-of-flow meter

27. Which of the following is **not** a danger associated with properties that have fixed fire extinguishing systems? *(547)*

    A. Explosive oxygen buildup as a result of carbon dioxide activation

    B. Poor visibility

    C. Energized electrical equipment

    D. Potential environmental impact

28. Firefighter A says that crews responding to bulk transport vehicle incidents must know the status and limitations of their water supply.

    Firefighter B says that crews responding to bulk transport vehicle incidents must determine the exact nature of the cargo as soon as possible.

    Who is right? *(533)*

    A. Firefighter A

    B. Firefighter B

    C. Both A and B

    D. Neither A nor B

29. Which of the following statements is correct? *(529)*

    A. Gasoline and kerosene are examples of flammable liquids.

    B. Gasoline and kerosene are examples of combustible liquids.

    C. Kerosene and vegetable oil are examples of combustible liquids.

    D. Kerosene and vegetable oil are examples of flammable liquids.

30. A major distinction between hydrocarbons and polar solvents is that ___ . *(529)*

    A. Hydrocarbons can be extinguished by foam, and polar solvents cannot

    B. Polar solvents can be extinguished by foam, and hydrocarbons cannot

    C. Hydrocarbons will mix with water, and polar solvents will not

    D. Polar solvents will mix with water, and hydrocarbons will not

31. Which of the following is **not** a similarity between fires involving transported liquid fuels and fires in liquid fuel storage areas? *(532)*

    A. Difficulties as a result of the amount of fuel

    B. The possibility of vessel failure

    C. Dangers to exposures

    D. Difficulty in containing spills and runoff

32. The response team for an incident involving a bulk transport vehicle should position themselves ___ as much as possible. *(532, 533)*

    A. Uphill, upwind, and on the curbside

    B. Downhill, upwind, and on the street-side

    C. Uphill, downwind, and on the curbside

    D. Downhill, downwind, and on the street-side

33. What is the recommended protocol for firefighters involved in shutting off gas at a meter? *(535)*

A. Request the utility to shut off the gas at the main distribution center.

B. Do not shut off the gas if there is not a fire.

C. Advance the team with a medium-expansion foam applied to the team using the rain-down method.

D. Advance the team with a fog pattern hoseline.

REVIEW TEST ANSWER SHEET

	A	B	C	D
1.	○	○	○	○
2.	○	○	○	○
3.	○	○	○	○
4.	○	○	○	○
5.	○	○	○	○
6.	○	○	○	○
7.	○	○	○	○
8.	○	○	○	○
9.	○	○	○	○
10.	○	○	○	○
11.	○	○	○	○
12.	○	○	○	○
13.	○	○	○	○
14.	○	○	○	○
15.	○	○	○	○
16.	○	○	○	○
17.	○	○	○	○
18.	○	○	○	○
19.	○	○	○	○
20.	○	○	○	○
21.	○	○	○	○
22.	○	○	○	○
23.	○	○	○	○
24.	○	○	○	○
25.	○	○	○	○
26.	○	○	○	○
27.	○	○	○	○
28.	○	○	○	○
29.	○	○	○	○
30.	○	○	○	○
31.	○	○	○	○
32.	○	○	○	○
33.	○	○	○	○

	A	B	C	D
34.	○	○	○	○
35.	○	○	○	○
36.	○	○	○	○
37.	○	○	○	○
38.	○	○	○	○
39.	○	○	○	○
40.	○	○	○	○
41.	○	○	○	○
42.	○	○	○	○
43.	○	○	○	○
44.	○	○	○	○
45.	○	○	○	○
46.	○	○	○	○
47.	○	○	○	○
48.	○	○	○	○
49.	○	○	○	○
50.	○	○	○	○
51.	○	○	○	○
52.	○	○	○	○
53.	○	○	○	○
54.	○	○	○	○
55.	○	○	○	○
56.	○	○	○	○
57.	○	○	○	○
58.	○	○	○	○
59.	○	○	○	○
60.	○	○	○	○
61.	○	○	○	○
62.	○	○	○	○
63.	○	○	○	○
64.	○	○	○	○
65.	○	○	○	○
66.	○	○	○	○
67.	○	○	○	○

	A	B	C	D
68.	○	○	○	○
69.	○	○	○	○
70.	○	○	○	○
71.	○	○	○	○
72.	○	○	○	○
73.	○	○	○	○
74.	○	○	○	○
75.	○	○	○	○
76.	○	○	○	○
77.	○	○	○	○
78.	○	○	○	○
79.	○	○	○	○
80.	○	○	○	○
81.	○	○	○	○
82.	○	○	○	○
83.	○	○	○	○
84.	○	○	○	○
85.	○	○	○	○
86.	○	○	○	○
87.	○	○	○	○
88.	○	○	○	○
89.	○	○	○	○
90.	○	○	○	○
91.	○	○	○	○
92.	○	○	○	○
93.	○	○	○	○
94.	○	○	○	○
95.	○	○	○	○
96.	○	○	○	○
97.	○	○	○	○
98.	○	○	○	○
99.	○	○	○	○
100.	○	○	○	○

Name _____

Date _____
Score _____

Chapter 14 Competency Profile

Student Name _____ Soc. Sec. No. _____

Last   First   Middle

Fire Department _____

Address _____

Phone _____

Home Address _____

Phone _____

Date of Enrollment ____ - ____ - ____   Total Class Hours _____

Date of Withdrawal ____ - ____ - ____   Total Hours Absent _____

Date of Completion ____ - ____ - ____

Instructor's Name _____   Session Dates _____

Instructor's Directions

1. Check the candidate's competency rating (3, 2, 1, ☒) for each performance test task and psychomotor lesson objective (practical activity and job sheets) listed below.

2. List any additional performance tasks or psychomotor objectives (job sheets or practical activity sheets) under "Other," and check the candidate's competency rating.

3. Record the candidate's cognitive scores (written lesson tests and *administered* chapter review tests) in the spaces provided.

Level				Psychomotor Competencies
3	2	1	☒	

Practical Activity Sheets

☐ ☐ ☐ ☐   PAS 14-1 — Determine Actions to Take, Including Retreat, When Dealing with Specific Conditons

☐ ☐ ☐ ☐   Other _____

☐ ☐ ☐ ☐   _____

Job Sheets

☐ ☐ ☐ ☐   JS 14-1 — Control and/or Extinguish an Exterior Class B Fire in an Open Pan

☐ ☐ ☐ ☐   Other _____

☐ ☐ ☐ ☐   _____

Level			
3	2	1	☒

Psychomotor Competencies

Chapter 14 Performance Test

☐	☐	☐	☐

Task 1 — Extinguish an ignitable liquid fire operating as a member of a team.

☐	☐	☐	☐

Task 2 — Control a flammable gas cylinder fire operating as a member of a team.

☐	☐	☐	☐
☐	☐	☐	☐

Other _____

_____

Points Achieved	Points Needed/ Total

Cognitive Competencies

Written Test

Points Achieved	Points Needed/Total	
_____	8/10	1. Distinguish between flammable liquids and combustible liquids.
_____	15/18	2. Select facts about suppressing Class B fires.
_____	3/4	3. Describe signs and effects of BLEVE.
_____	4/4	4. List the four ways that water can be used to attack a Class B fire.
_____	4/5	5. List methods of identifying tank contents.
_____	4/5	6. Select facts about techniques for suppressing bulk transport vehicle fires.
		7. Evaluated on Job Sheet 14-1
_____	12/15	8. Distinguish between the characteristics of natural gas and liquid petroleum gas.
		9. Evaluated on Practical Activity Sheet 14-1

Review Test

_____ Chapter 14 Review Test

Instructor's Signature _____ **Date** _____

Student's Signature _____ **Date** _____

STUDENT APPLICATIONS

FOURTH EDITION

ESSENTIALS OF FIRE FIGHTING

LESSON
15

FIRE DETECTION, ALARM, & SUPPRESSION SYSTEMS

FIREFIGHTER II

FIRE PROTECTION PUBLICATIONS
OKLAHOMA STATE UNIVERSITY

Study Objectives

LESSON OBJECTIVE

After completing this lesson, you will be able to discuss the operation of typical automatic fire detection and suppression systems. You will also be able to identify the components of typical automatic sprinkler systems and to inspect those systems.

ENABLING OBJECTIVES

After reading Chapter 15 of *Essentials,* pages 559 through 571 and 578 through 584, and completing related activities, you will be able to —

1. Match types of alarm-initiating devices to their descriptions.

2. Select facts about heat detectors.

3. Select facts about smoke detectors.

4. Complete statements about flame detectors.

5. Complete statements about fire-gas detectors.

6. State the reason for having a variety of alarm-indicating devices.

7. Match types of automatic alarm systems to their descriptions.

8. Select facts about supervising fire alarm systems.

9. List auxiliary services provided by fire detection and alarm systems.

10. Complete statements about water flow alarms.

11. Match sprinkler system applications to their descriptions.

12. Identify components of fire suppression systems.

13. **Inspect protected property fire suppression systems.** *(Job Sheet 15-1)*

Study Sheet

Introduction

This study sheet is intended to help you learn the Firefighter II material in Chapter 15 of *Essentials of Fire Fighting,* Fourth Edition. You may use it for self-study, or you may use it to review material that will be covered in the lesson and chapter review tests. The numbers in parentheses are the pages in *Essentials* on which the answers or terms can be found.

Chapter Vocabulary

Be sure that you know the chapter-related meanings of the following terms and abbreviations. Use a dictionary or the glossary in *Fire Service Orientation and Terminology* if you cannot determine the meaning of the term from its context.

- Bimetallic detector *(561)*
- Continuous-line detection device *(561)*
- Central station *(569)*
- Frangible bulb *(560)*
- Fusible device *(560)*
- Local energy system *(567)*

- Parallel telephone system *(568)*
- Photoelectric detector *(563)*
- Proprietary system *(569)*
- Remote station *(568)*
- Shunt system *(568)*
- Spot-type detector *(561)*

Study Questions & Activities

1. What are the five recognized functions of fire detection and alarm systems? *(559)*

   a. _____

   _____

   b. _____

   _____

   c. _____

   _____

   d. _____

   _____

   e. _____

   _____

2. What is the most reliable form of fixed fire protection system? *(559)*

   _____

3. How does a local system or protected premises fire alarm system work? *(559)*

   _____

   _____

   _____

   _____

4. What are the four basic types of automatic alarm-initiating devices? *(560)*

   a. _____

   b. _____

   c. _____

   d. _____

5. Distinguish between fixed-temperature and rate-of-rise heat detectors. *(560–563)*

   a. Fixed-temperature _____

   _____

   b. Rate-of-rise _____

   _____

6. Discuss the recommended location and sensitivity of a heat detector. *(560)*

   a. Location _____

   _____

   b. Sensitivity _____

   _____

7. What are the three primary principles of physics by which heat detectors work? *(560)*

   a. _____

   b. _____

   c. _____

8. What are the two basic types of continuous-line heat detectors? *(561)*

   a. _____

   b. _____

9. Describe the operation of a bimetallic detector. *(562)*

   _____

   _____

   _____

   _____

10. What are the limitations in length and spacing of pneumatic rate-of-rise line detectors? *(562)*

   _____

   _____

   _____

11. What type of rate-of-rise detector is recommended for use in areas that undergo regular temperature changes? *(563)*

   _____

   _____

12. Why are smoke detectors preferable to heat detectors in many types of occupancies? *(562)*

_____

_____

13. What are the two basic types of smoke detectors? *(563)*

   a. _____

   b. _____

14. Describe the operation of the two types of photoelectric detectors. *(563, 564)*

   a. Beam application_____

      _____

   b. Refractory photocell _____

      _____

15. Describe the operation of an ionization smoke detector. *(564)*

_____

_____

_____

16. What are the three types of flame detectors? *(565)*

   a. _____

   b. _____

   c. _____

17. What are the two gases normally monitored by fire-gas detectors? *(566)*

   a. _____

   b. _____

18. State at least two reasons for having various types of indicators for alarm systems. *(566)*

   a. _____

      _____

   b. _____

      _____

19. What are the three basic types of auxiliary systems? *(567)*

   a. _____

   b. _____

   c. _____

20. What distinguishes a remote station system from an auxiliary system? *(568)*

_____

_____

_____

21. What are the functions and common capabilities of proprietary systems? *(569)*

_____

_____

_____

_____

_____

22. What is the principal difference between a central station and a proprietary system? *(569)*

_____

_____

_____

_____

23. What is meant by the term *self-supervising* as applied to an alarm system? *(570)*

_____

_____

_____

_____

24. What methods are used to protect people when systems such as a CO_2 system are activated? *(570)*

_____

_____

_____

_____

25. What are the two types of water flow alarms? *(578)*

    a. _____

    b. _____

26. List the five major types of sprinkler systems in use today. *(580)*

    a. _____

    b. _____

    c. _____

    d. _____

    e. _____

27. Compare the wet-pipe automatic sprinkler system and the dry-pipe automatic sprinkler system. *(580, 581)*

    a. Wet-pipe system _____

    _____

    _____

    b. Dry-pipe system_____

_____

_____

28. What is the purpose of a retard chamber in a wet-pipe sprinkler system? How does it function? *(580)*

_____

_____

_____

_____

29. What are the air pressure requirements for dry systems? How is this air pressure derived? *(581)*

_____

_____

_____

30. What is the purpose of an accelerator? How does it work? *(581)*

_____

_____

_____

_____

31. What is a preaction system and when is it used? *(581)*

_____

_____

_____

_____

32. Where would deluge sprinkler systems most likely be installed? Why? *(581, 582)*

_____

_____

_____

33. What are some different modes of detection in a deluge system? *(582)*

_____

_____

_____

34. What are the various ways of operating the deluge valves of a deluge system? *(582)*

_____

_____

_____

35. What are the main purposes of residential sprinkler systems? *(582)*

_____

_____

_____

36. Describe the features of residential sprinkler systems. *(582, 583)*

_____

_____

_____

_____

Job Sheet 15-1
Inspect Protected Property Fire
Suppression Systems

Name _____ Date _____

Evaluator _____ Overall Competency Rating _____

References	NFPA 1001, Prevention, Preparedness, and Maintenance 4-3.4a ***Essentials,*** pages 571–583
Prerequisites	None
Student's Instructions	To meet evaluation standards, you must perform this job within _____ *[amount of time, if applicable];* you may have _____ attempts. When you are ready to perform this job, ask your instructor to observe the procedure and complete this form. To show mastery of this job, you must perform all steps to receive an overall competency rating of at least 2.

<table>
<tr><td>

Competency Rating Scale

3 — Skilled — Meets all evaluation criteria and standards; performs task independently on first attempt; requires no additional practice or training.

2 — Moderately skilled — Meets all evaluation criteria and standards; performs task independently; additional practice is recommended.

1 — Unskilled — Is unable to perform the task; additional training required.

☒ — **Unassigned** — Job sheet task is not required or has not been performed.

✔ **Evaluator's Note:** Formulate and inform the candidate of the standards for this task (time allowed and number of attempts). Observe the candidate perform the task, check the step/key point under the appropriate attempt number as accomplished, record total time (if appropriate), and then use the rating scale above to assign an overall competency rating. If the candidate is unable to perform any step of this job, have the candidate review the materials and try again.

</td></tr>
</table>

Introduction	Firefighters should pay particular attention to the fire protection systems in the occupancies they inspect. Firefighters will almost never actually test the systems themselves; however, regardless of whether or not the system is tested, firefighters should inspect each of the listed systems and note any deficiencies on the inspection form.
Equipment and Personnel	• Inspection team in uniform dress • Clipboard • Inspection form • Writing implement

Job Steps	Key Points	Attempt No.		
		1	2	3

FIRE DETECTION AND ALARM SYSTEMS

1. Inspect all fire detection and alarm systems.

   1. a. Recording any that are inoperable   ___ ___ ___

       b. Recording any missing system components (e.g., detector heads)   ___ ___ ___

2. Inspect the annunciator panels.

   2. a. Recording status of power supply (should be on)   ___ ___ ___

       b. Recording mode (should not be in trouble mode)   ___ ___ ___

Time (Total)   ___ ___ ___

Evaluator's Comments _____

_____

_____

FIXED EXTINGUISHING SYSTEMS

1. Inspect the water supply valves.

   1. Recording on inspection form any open valves (except OS&Y, which should be locked in open position)   ___ ___ ___

2. Inspect the pressure gauges.

   2. Recording pressures on inspection form   ___ ___ ___

3. Inspect the system's components.

   3. Recording on inspection form any damage or signs of tampering   ___ ___ ___

4. Inspect the sprinkler heads.

   4. Recording on inspection form any that have been painted, damaged, or obstructed   ___ ___ ___

5. Inspect special agent extinguishing systems.

   5. Recording on inspection form any that are not charged and ready for service   ___ ___ ___

Time (Total)   ___ ___ ___

Evaluator's Comments _____

_____

_____

Chapter 15 Review Test

> ➡ **Directions:** This review test covers the Firefighter II material in Chapter 12 of your *Essentials of Fire Fighting* text. It may be assigned as a study aid (self-test) or may be administered by your instructor as a pretest or posttest.
>
> When used as a study aid, try to answer the questions without referring to the page numbers in *Essentials* or your *Firefighter II Student Applications* workbook *(SA)* on which the answers can be found until after you have completed the entire test. Then check your answers against those on the pages provided in parentheses.
>
> When administered by your instructor as a pretest or posttest, read each of the test questions carefully. Choose the best response and then darken the corresponding letter on your answer sheet.
>
> This chapter review test contains 50 multiple-choice questions, each worth 2 points. To pass the test, you must achieve at least 84 of the 100 points possible.

1. What is a protected premises fire alarm system? *(559)*

    A. Hydraulic or pneumatic system that notifies the local jurisdiction

    B. Hydraulic or pneumatic system that notifies only those in the facility

    C. Manual or automatic system that notifies the local jurisdiction

    D. Manual or automatic system that notifies only those in the facility

2. Which of the following statements is true of fixed-temperature heat detectors? *(560)*

    A. They should be placed in the upper portions of the room such as the ceiling.

    B. They are very prone to false alarms.

    C. They should have an activation temperature slightly below the highest temperature expected in the room where installed.

    D. They provide quick response to high temperatures.

3. What is the basis of operation for a fusible device? *(560)*

    A. At a given temperature, a wire breaks to allow a plunger to fall and complete an alarm circuit.

    B. At a given temperature, two solder contacts melt, flow together, and complete an alarm circuit.

    C. At a given temperature, a glass vial breaks to allow two contacts to come together to complete an alarm circuit.

    D. At a given temperature, solder melts to allow a plunger to fall and complete an alarm circuit.

4. Firefighter A says that heat detectors that continuously monitor heat in a single location are referred to as continuous-line detectors.

   Firefighter B says that heat detectors that monitor heat in a number of spots throughout a facility are referred to as spot detectors.

   Who is right? *(561)*

   A. Firefighter A   C. Both A and B

   B. Firefighter B   D. Neither A nor B

5. What is the basis of operation for a frangible bulb in a detection device? *(560)*

   A. At a given temperature, a wire breaks to allow a plunger to fall and complete an alarm circuit.

   B. At a given temperature, two solder contacts melt, flow together, and complete an alarm circuit.

   C. At a given temperature, a glass vial breaks to allow two contacts to come together to complete an alarm circuit.

   D. At a given temperature, solder melts to allow a plunger to fall and complete an alarm circuit.

6. What is the function of the thermistor core in a tubing-type continuous line detector? *(561)*

   A. Serves as insulation between two conductive surfaces and melts when the temperature gets high enough so that the conductors make contact

   B. Expands in the presence of increased heat so that contacts at each end of the detector complete the alarm circuit

   C. Provides insulation between the two conductive layers until increased heat changes its resistance so that more current flows between the conductive layers

   D. Conducts current between two layers of insulation; the current flow increases as temperature increases until the insulation melts

7. What is the function of the heat-sensitive material in a wire-type continuous line detector? *(561)*

   A. Serves as insulation between two conductive surfaces and melts when the temperature gets high enough so that the conductors make contact

   B. Expands in the presence of increased heat so that contacts at each end of the detector complete the alarm circuit

   C. Provides insulation between the two conductive layers until increased heat changes its resistance so that more current flows between the conductive layers

   D. Conducts current between two layers of insulation; the current flow increases as temperature increases until the insulation melts

8. What is the principle of operation of rate-of-rise heat detectors? *(562)*

   A. The system compares the temperature at a location at regular intervals, such as every five minutes, and activates if there is a significant increase.

   B. The system activates an alarm if the temperature increases more than a specified amount in one minute.

C. The system is actuated by rising vapors and gases with higher than normal temperatures.

D. The system activates an alarm if the temperature rises above a certain level and remains there for at least five minutes or continues to rise after five minutes.

9. What is the principle of operation for bimetallic detectors? *(561, 562)*

A. An actuator bends because the two metals that compose it expand at different rates when exposed to heat.

B. Two metal contacts bend toward each other, due to a combination of heat and magnetism, until they complete an alarm circuit.

C. A nonconductive actuator expands at a controlled rate until it completes an alarm circuit.

D. Two metal alloy contacts expand at a controlled rate until they touch to close a circuit switch.

10. Firefighter A says that pneumatic heat detectors are powered by air compressors for line units and sealed bladders for spot units.

Firefighter B says that pneumatic line detectors are limited to 100 feet *(30 m)* of total length in rows spaced no more than 6 feet *(2 m)* apart.

Who is right? *(562)*

A. Firefighter A

B. Firefighter B

C. Both A and B

D. Neither A nor B

11. What type of rate-of-rise detector is used in areas that are normally subject to regular temperature changes at temperatures lower than would be present during a fire? *(563)*

A. Pneumatic spot

B. Pneumatic line

C. Thermoelectric

D. Rate-compensated

12. What type of smoke detector uses an external light source and a receiving unit? *(563, 564)*

A. Ionization

B. Refractory photocell

C. Beam-application photoelectric

D. Photo refraction

13. What type of smoke detector uses an internal light source and photocell? *(563, 564)*

A. Ionization

B. Refractory photocell

C. Beam-application photoelectric

D. Photo refraction

14. How is the alarm activated in an ionization smoke detector? *(564)*

A. Ionized smoke particles within the detector chamber cause an increased flow of electrical current.

B. Ionized smoke particles within the detector chamber cause a decreased flow of electrical current.

C. Ionized light flows more quickly between the light sender and detector.

D. Ionized light flows at a slower rate between the light sender and detector.

15. Which type of smoke detector responds faster to smoldering fires? *(564)*

   A. Ionization

   B. Photoelectric

   C. Ultraviolet

   D. Infrared

16. What is the recommended interval for changing smoke detector batteries? *(565)*

   A. When they fail an annual battery test

   B. Annually, whether or not it fails a battery test

   C. Every six months

   D. Monthly

17. Firefighter A says that flame detector systems should be installed in sheltered, out-of-the-way places where they are protected from direct light, such as behind furniture, to avoid false alarms.

   Firefighter B says that some flame detectors are sensitive to both ultraviolet and infrared portions of the light spectrum and that some are sensitive only to flickering light.

   Who is right? *(565)*

   A. Firefighter A

   B. Firefighter B

   C. Both A and B

   D. Neither A nor B

18. Which of the following is an environment where ultraviolet flame detectors are suited for use? *(565, 566)*

   A. Areas where welding is done

   B. Areas exposed to bright sunlight

   C. Areas that are illuminated by intense mercury-vapor lamps

   D. Areas where infrared flame detectors are in use

19. Firefighter A says that fire-gas detectors actuate alarms faster than smoke detectors but slower than heat detectors.

   Firefighter B says that water vapor, carbon dioxide, and carbon monoxide are the only gases released from all fires.

   Who is right? *(566)*

   A. Firefighter A

   B. Firefighter B

   C. Both A and B

   D. Neither A nor B

20. What is the most valuable use of gas detectors? *(566)*

   A. Discriminating between friendly and hostile fires

   B. Identifying whether a fire contains toxic gases

   C. Providing a quick alert of a fire presence

   D. Verifying alarms provided by flame detector systems

21. What is the function of a local energy system? *(567)*

    A. Notifying personnel within a facility

    B. Notifying personnel within a facility and signaling the local jurisdiction

    C. Signaling the local jurisdiction without sounding an alarm in the facility

    D. Notifying passersby

22. Why have a variety of alarm-indicating devices been developed? *(566)*

    A. To accommodate those with vision or hearing impairments

    B. To provide alarms in work areas where personnel wear noise-attenuation devices

    C. To provide an alarm distinct from other alarms and noises employed in a facility

    D. All of the above

23. What type of system uses an extension circuit from the municipal system into a protected facility? *(568)*

    A. Local energy system

    B. Shunt system

    C. Parallel telephone system

    D. Remote station system

24. Which of the following statements is true with regard to remote station systems? *(568)*

    A. A coded alarm system is used for single-occupancy properties.

    B. A noncoded alarm system is used for multiple-occupancy properties.

    C. They must have a means of notifying the fire alarm center when the system becomes impaired.

    D. They are connected to the municipal fire-alarm-box system.

25. What is the purpose of a proprietary alarm system? *(569)*

    A. Monitoring a protected property and notifying a commercial protection company of an emergency

    B. Monitoring the apartments within a building

    C. Monitoring a number of facilities, such as a college campus, through a staffed receiving station, which notifies jurisdiction resources

    D. Monitoring particular hazards such as nuclear plants

26. What is a central station system? *(570)*

    A. Contracted service that monitors systems for individual customers

    B. Main dispatch office for a municipal jurisdiction

    C. Automated dispatching center that transmits alarm to various types of emergency services

    D. 9-1-1 dispatch facility

27. What are the auxiliary systems related to fire detection and alarm systems? *(571)*

    A. The power source and alarm test system

    B. The fasteners and related wiring and switches required for the installation

    C. Systems that integrate process and environmental controls, security, and personnel-access controls

    D. Interfaces between fire detection and fire suppression systems

28. Firefighter A says that the biggest problem with fire alarm systems is that they do not offer a means of self-monitoring or self-diagnosis in the event of problems.

    Firefighter B says that fire alarm systems would be more functional if they could provide protection; for example, CO_2 systems should notify personnel of oxygen-deficient atmospheres.

    Who is right? *(570)*

    A. Firefighter A   C. Both A and B

    B. Firefighter B   D. Neither A nor B

29. What are the two most common types of water flow alarms? *(578)*

    A. Hydraulic and electric   C. Electric and pneumatic

    B. Hydraulic and pneumatic   D. Wet and dry

30. What activates water flow alarms? *(578)*

    A. The person who activates the sprinkler system

    B. Any person who notices that water is flowing through the sprinkler system

    C. The flow of water in the fire department connection

    D. The flow of water in the sprinkler system

31. Which of the following detectors does ***not*** reset automatically? *(560–563)*

    A. Pneumatic rate-of-rise spot heat detector

    B. Fusible-device fixed-temperature heat detector

    C. Bimetallic detector

    D. Rate compensated detector

32. Which of the following is an example of a combination detector? *(566)*

    A. Fixed-temperature/rate-of-rise heat detectors

    B. Heat/smoke detectors

    C. Smoke/fire-gas detectors

    D. All of the above

33. Which of the following is ***not*** one of the principles of physics used to detect heat in fixed-temperature devices? *(560)*

    A. Expansion of heated material

    B. Changes in the electrical voltage of heated material

    C. Melting of heated material

    D. Changes in electrical resistance of heated material

34. What is the simplest type of sprinkler system? *(580)*

    A. Deluge
    B. Dry-pipe
    C. Wet-pipe
    D. Preaction

35. What is the purpose of a retard chamber in a wet-pipe sprinkler? *(580)*

    A. Reducing the pressure flow through the system
    B. Reducing the chances of the system freezing
    C. Reducing the chances for false alarm activation
    D. Preventing water from flowing backwards through the system

36. Firefighter A says that wet-pipe sprinkler systems use pipes set at an angle to allow water to drain to prevent freezing.

    Firefighter B says that wet-pipe sprinkler systems contain water under pressure at all times.

    Who is right? *(580)*

    A. Firefighter A
    B. Firefighter B
    C. Both A and B
    D. Neither A nor B

37. What medium is used to keep water out of dry-pipe systems until activated? *(580)*

    A. Mechanical valves
    B. Hydraulic pistons
    C. Air under pressure
    D. Spring-loaded check valves

38. Quick-opening devices are required on dry-pipe sprinkler systems with a water capacity of more than ___. *(581)*

    A. 100 gallons *(400 L)*
    B. 250 gallons *(1 000 L)*
    C. 500 gallons *(2 000 L)*
    D. 1,000 gallons *(4 000 L)*

39. What are the components of a preaction sprinkler system? *(581)*

    A. Drip valve, manual alarm system, and sprinklers
    B. Retard valve, fire detection devices, and hoses
    C. Dry-pipe valve, manual alarm system, and sprinklers
    D. Deluge-type valve, fire detection devices, and sprinklers

40. What type of actuation system is used with a deluge sprinkler system? *(582)*

    A. Sensors in each sprinkler head
    B. Manual actuation valves
    C. A separate detection system
    D. Sensors at each pipe connection

41. What is the purpose of a residential sprinkler system? *(582)*

    A. To prevent total fire involvement in the room of origin and to give occupants of the dwelling a chance to escape

    B. To prevent total fire involvement throughout the residence and to give occupants of the dwelling a chance to escape

    C. To prevent total fire involvement in the room of origin and to notify the fire department

    D. To prevent total fire involvement throughout the residence and to notify the fire department

42. Which of the following statements is true with regard to residential sprinkler systems? *(583)*

    A. The system may be connected directly to the public water supply or tied to the dwelling's domestic water system.

    B. Residential water systems cannot be equipped with control valves that allow them to be isolated from the water supply for safety reasons.

    C. Residential systems can only be of the wet-pipe design.

    D. Residential systems cannot be equipped with fire department connections due to the high pressures produced by pumpers.

43. Which of the following devices is **not** required for a residential sprinkler system? *(582, 583)*

    A. Pressure gauge

    B. System testing device

    C. Flow detector

    D. Ball drip valve

44. What NFPA standard addresses the primary guidelines for design and installation of sprinkler systems? *(571)*

    A. 1910   C. 13

    B. 14A   D. 1970

45. Which sprinkler valve is usually located in a remote part of the sprinkler system and is used to simulate the activation of one sprinkler? *(578)*

    A. Post indicator valve

    B. Wall post indicator valve

    C. Inspector's test valve

    D. Outside screw and yoke valve

46. What is the required air pressure for dry-pipe systems? *(581)*

    A. 10 psi *(70 kPa)* below the trip pressure

    B. 20 psi *(140 kPa)* below the trip pressure

    C. 20 psi *(140 kPa)* above the trip pressure

    D. 15 psi *(100 kPa)* above the trip pressure

47. What valve discharges water from every open head connected to the system controlled by that specific valve? *(582)*

    A. OS&Y

    B. Deluge

    C. PIVA

    D. Ball drip

48. What NFPA standard covers residential sprinkler systems? *(582)*

    A. 13D

    B. 1910

    C. 14A

    D. 1907

49. For what reason is a sprinkler system least likely to fail? *(572)*

    A. Interruption to the municipal water supply

    B. Failure of the actual sprinklers

    C. Frozen or broken pipes

    D. Excess debris or sediment in the pipes

50. Where would a deluge system most likely be found? *(582)*

    A. Residence

    B. Computer lab

    C. Aircraft hangar

    D. Post office

REVIEW TEST ANSWER SHEET

	A	B	C	D
1.	○	○	○	○
2.	○	○	○	○
3.	○	○	○	○
4.	○	○	○	○
5.	○	○	○	○
6.	○	○	○	○
7.	○	○	○	○
8.	○	○	○	○
9.	○	○	○	○
10.	○	○	○	○
11.	○	○	○	○
12.	○	○	○	○
13.	○	○	○	○
14.	○	○	○	○
15.	○	○	○	○
16.	○	○	○	○
17.	○	○	○	○
18.	○	○	○	○
19.	○	○	○	○
20.	○	○	○	○
21.	○	○	○	○
22.	○	○	○	○
23.	○	○	○	○
24.	○	○	○	○
25.	○	○	○	○
26.	○	○	○	○
27.	○	○	○	○
28.	○	○	○	○
29.	○	○	○	○
30.	○	○	○	○
31.	○	○	○	○
32.	○	○	○	○
33.	○	○	○	○

	A	B	C	D
34.	○	○	○	○
35.	○	○	○	○
36.	○	○	○	○
37.	○	○	○	○
38.	○	○	○	○
39.	○	○	○	○
40.	○	○	○	○
41.	○	○	○	○
42.	○	○	○	○
43.	○	○	○	○
44.	○	○	○	○
45.	○	○	○	○
46.	○	○	○	○
47.	○	○	○	○
48.	○	○	○	○
49.	○	○	○	○
50.	○	○	○	○
51.	○	○	○	○
52.	○	○	○	○
53.	○	○	○	○
54.	○	○	○	○
55.	○	○	○	○
56.	○	○	○	○
57.	○	○	○	○
58.	○	○	○	○
59.	○	○	○	○
60.	○	○	○	○
61.	○	○	○	○
62.	○	○	○	○
63.	○	○	○	○
64.	○	○	○	○
65.	○	○	○	○
66.	○	○	○	○
67.	○	○	○	○

	A	B	C	D
68.	○	○	○	○
69.	○	○	○	○
70.	○	○	○	○
71.	○	○	○	○
72.	○	○	○	○
73.	○	○	○	○
74.	○	○	○	○
75.	○	○	○	○
76.	○	○	○	○
77.	○	○	○	○
78.	○	○	○	○
79.	○	○	○	○
80.	○	○	○	○
81.	○	○	○	○
82.	○	○	○	○
83.	○	○	○	○
84.	○	○	○	○
85.	○	○	○	○
86.	○	○	○	○
87.	○	○	○	○
88.	○	○	○	○
89.	○	○	○	○
90.	○	○	○	○
91.	○	○	○	○
92.	○	○	○	○
93.	○	○	○	○
94.	○	○	○	○
95.	○	○	○	○
96.	○	○	○	○
97.	○	○	○	○
98.	○	○	○	○
99.	○	○	○	○
100.	○	○	○	○

Name _____

Date _____

Score _____

Chapter 15 Competency Profile

Student Name _____ Soc. Sec. No. _____

Last   First   Middle

Fire Department _____

Address _____

Phone _____

Home Address _____

Phone _____

Date of Enrollment _____ - _____ - _____   Total Class Hours _____

Date of Withdrawal _____ - _____ - _____   Total Hours Absent _____

Date of Completion _____ - _____ - _____

Instructor's Name _____   Session Dates _____

Instructor's Directions

1. Check the candidate's competency rating (3, 2, 1, ☒) for each performance test task and psychomotor lesson objective (practical activity and job sheets) listed below.

2. List any additional performance tasks or psychomotor objectives (job sheets or practical activity sheets) under "Other," and check the candidate's competency rating.

3. Record the candidate's cognitive scores (written lesson tests and *administered* chapter review tests) in the spaces provided.

Level				Psychomotor Competencies
3	2	1	☒	

Practical Activity Sheets

None Required

Other _____

☐ ☐ ☐ ☐

☐ ☐ ☐ ☐

Job Sheets

JS 15-1 — Inspect Protected Property Fire Suppression Systems

Other _____

☐ ☐ ☐ ☐

☐ ☐ ☐ ☐

☐ ☐ ☐ ☐

Level
3 2 1 ☒

☐	☐	☐	☐
☐	☐	☐	☐
☐	☐	☐	☐

Psychomotor Competencies

Chapter 15 Performance Test

Task 1 — Connect hoseline to the fire department connection (FDC) of a sprinkler system, interpret pressure gauges, and operate system control valves.

Other _____

_____

Points Achieved	Points Needed/ Total

Cognitive Competencies

Written Test

Points Achieved	Points Needed/Total	
_____	9/11	1. Match types of alarm-initiating devices to their descriptions.
_____	6/7	2. Select facts about heat detectors.
_____	5/6	3. Select facts about smoke detectors.
_____	3/4	4. Complete statements about flame detectors.
_____	3/4	5. Complete statements about fire-gas detectors.
_____	2/2	6. State the reason for having a variety of alarm-indicating devices.
_____	5/6	7. Match types of automatic alarm systems to their descriptions.
_____	4/5	8. Select facts about supervising fire alarm systems.
_____	8/8	9. List auxiliary services provided by fire detection and alarm systems.
_____	4/4	10. Complete statements about water flow alarms.
_____	4/5	11. Match sprinkler system applications to their descriptions.
_____	8/10	12. Identify components of fire suppression systems.
		13. Evaluated on Job Sheet 15-1

Review Test

Chapter 15 Review Test

_____

Instructor's Signature _____ **Date**_____

Student's Signature _____ **Date**_____

STUDENT APPLICATIONS

FOURTH EDITION
ESSENTIALS OF FIRE FIGHTING

LESSON 17

FIRE CAUSE & ORIGIN

FIREFIGHTER II

FIRE PROTECTION PUBLICATIONS
OKLAHOMA STATE UNIVERSITY

Study Objectives

LESSON OBJECTIVE

After completing this lesson, you will be able to identify your responsibilities in fire cause determination and protect evidence of fire cause and origin.

ENABLING OBJECTIVES

After reading Chapter 17 of *Essentials* pages 621, 622, and 627 through 629 and completing related activities, you will be able to —

1. List responsibilities of a fire investigator.

2. Select facts about conduct and statements at the scene.

3. Select facts about securing the scene and legal considerations.

4. Select facts about protecting and preserving evidence.

5. **Protect evidence of fire cause and origin. *(Job Sheet 17-1)***

6. **Assess the origins and causes of fires. *(Practical Activity Sheet 17-1)***

Study Sheet

Introduction | This study sheet is intended to help you learn the material in Chapter 17 of ***Essentials of Fire Fighting,*** Fourth Edition. You may use it for self-study, or you may use it to review material that will be covered in the lesson and chapter review tests. The numbers in parentheses are the pages in ***Essentials*** on which the answers or terms can be found.

Chapter Vocabulary | Be sure that you know the chapter-related meanings of the following terms and abbreviations. Use a dictionary or the glossary in ***Fire Service Orientation and Terminology*** if you cannot determine the meaning of the term from its context.

- Michigan *vs.* Tyler case *(628)*

- Search warrant *(628)*

Study Questions & Activities

1. List the factors that combine to cause a fire. *(621)*

   a. _____

   b. _____

   c. _____

2. What is the purpose of analyzing the causes of fire? *(621)*

   _____

   _____

3. In most jurisdictions, who has the ultimate legal responsibility for fire cause determination? *(622)*

   _____

4. Why do firefighters have the important responsibility of noting everything that could point to the cause of the fire? *(622)*

   _____

   _____

5. What four questions should the first-arriving firefighters ask with regard to fire cause? *(622)*

   a. _____

   b. _____

   c. _____

   d. _____

6. Who actually carries out the duties of fire cause determination? *(622)*

   _____

   _____

7. When can information on fire cause be made public? *(627)*

    _____

    _____

8. What statement is a sufficient reply to any question by the public concerning fire cause? *(627)*

    _____

9. What is the purpose of properly securing the fire scene? *(628)*

    _____

    _____

10. What procedures should be followed if an investigator is not immediately available? *(628)*

    _____

    _____

    _____

    _____

11. What are the legal rights of the fire department at the scene of a fire? After the fire? *(628)*

    _____

    _____

    _____

    _____

12. List some ways the fire-scene premises can be secured and protected. *(628)*

    _____

    _____

    _____

    _____

13. If there is evidence of possible arson, why should at least one person remain on the premises until the investigator arrives? *(629)*

    _____

    _____

    _____

14. Where can the fire department learn the legal opinions that affect its jurisdiction in arson investigation? *(629)*

    _____

    _____

    _____

    _____

15. What is the general guideline for the handling of possible evidence by firefighters? *(629)*

_____

_____

_____

_____

16. What is meant by chain of custody? *(629)*

_____

_____

_____

17. What are your department's SOPs regarding arson investigation? *(Local protocol)*

_____

_____

_____

Information Sheet 17-1

Common Causes of Fires and Their Prevention

Category	Cause	Prevention
Mechanical failure	• Part failure or malfunction • Electrical short or ground fault	• Inspection of specific part that has a tendency to fail or wear • Routine inspection of parts
Misuse of heat	• Smoking materials • Children • Unconsciousness (sleeping, mental impairment, under the influence) • Welding operations • Open fires • Improper cooking habits	• Public inspection and education
Arson and incendiary acts	• Deliberately set fires • Fire bombs	• Investigation • Public education
Operational deficiencies in machines and systems	• Collision or knocked down • Accidental ON/OFF • Unattended • Overloaded • Design alteration • Animals	• Public education • Prevention inspection • Constructing and modifying equipment to keep animals away
Misuse of material ignited	• Accidentally released or spilled • Improperly stored • Improperly combined • Improper container • Accessible to children	• Proper handling • Public education • Learn Not to Burn® program

Information Sheet 17-1 (Continued)

Category	Cause	Prevention
Construction or installation deficiencies	• Hazardous design • Faulty construction • Improper installation	• Plan check • On-site inspections during construction • Building inspections • Home safety surveys
Nature	• Earthquake • High wind • Floods • Lightning	• No prevention • Public education and preplanning help lessen the problems
Rekindle	• Incomplete extinguishment	• Search for and extinguishment of hidden fire and hot spots • Total extinguishment

Information Sheet 17-2
Types of Evidence

EN ROUTE

Time of Day	Weather and Natural Hazards	Man-Made Barriers	People Leaving the Scene
Are people dressed appropriately for the time of day? (*Are they wearing night clothes at 3 a.m., work clothes during working hours, etc.?*)

Are people where they normally would be at this time of day? (*Are they home at 3 a.m., not at the office?*) | What is the weather? (*Arsonists sometimes set fires during inclement weather because the fire department's response time will be longer.*)

Is the furnace operating even though the outside temperature is high?

Are the windows wide open even though the outside temperature is low? | Are there any barriers (*barricades, fallen trees, cables, trash containers, vehicles*) blocking access to hydrants, sprinkler and standpipe connections, streets, or driveways? | Are people leaving the scene by vehicle or foot? (*Most people are intrigued by a fire and stay to watch.*) |

UPON ARRIVAL

Doors and Windows	Location of Fire	Containers or Cans	Burglary Tools	Familiar Faces
Are doors or windows locked or unlocked?

Do doors or windows show signs of forced entry (broken glass, split frames, etc.)?

Are doors or windows covered with blankets, paper, or paint? (*Sometimes an arsonist will cover windows to delay discovery of the fire.*) | What is the location of the fire? (*This information helps determine the area of origin.*)

Are there separate seemingly unconnected fires? (*If so, the fires may have been set in several locations or spread by trailers.*) | Are there any metal cans or plastic containers discarded inside or outside the structure? (*They may have been used to transport accelerants.*) | Are there any tools such as pry bars or screwdrivers in unusual areas? (*They may have been used by an arsonist to enter the facility.*) | Are there any familiar faces in the crowd of bystanders? (*They may be fire buffs or habitual firesetters.*) |

Information Sheet 17-2 (Continued)

DURING FIRE FIGHTING

Unusual Odors	Abnormal Fire Behavior	Obstacles Hindering Fire Fighting	Incendiary Devices	Trailers
Are there any unusual odors? (*Unusual odors may sometimes be detected at the fire scene even though firefighters are wearing SCBA.*)	Does the fire flashback, reignite, rekindle several times in the same area, or increase in intensity or spread in several directions when water is applied? (*These are signs of possible accelerant use.*)	Are doors tied shut or is furniture placed in doorways or hallways to hinder fire fighting efforts? Are holes cut in floors? (*This may not only hinder firefighters but spread the fire.*)	Are there any pieces of glass, fragments of bottles or containers, or metal parts of mechanical devices in evidence? (*Most incendiary devices leave evidence of their existence.*)	Are there combustible materials such as rolled rags, blankets, newspapers, or ignitable liquid that could have been used to spread the fire from one area to another? (*Trailers usually leave char or burn patterns and may be used with incendiary devices.*)

Structural Alterations	Fire Patterns	Heat Intensity	Availability of Documents	Fire Detection and Protection Systems
Has plaster or drywall been removed to expose wood? Have holes been made in the ceilings, walls, or floors? Are fire doors secured in an open position?	Are there areas of irregular burning or locally heavy charring in areas of little fuel? (*The fire's movement and intensity patterns can help trace how the fire spread.*)	Is there evidence of high heat intensity, especially in relation to other areas of the same room? (*This may indicate the use of accelerants.*)	Can the occupant suddenly produce insurance papers, inventory lists, deeds, or other legal documents? (*This may indicate that the fire was premeditated.*)	Have fire detection or protection systems been intentionally damaged or rendered inoperable? Have intrusion alarms been tampered with or intentionally disabled?

Location of Fire	Personal Possessions	Household Items	Equipment or Inventory	Business Records
Are there logical ignition sources in the area of the fire? (*Fires in closets, bathtubs, file drawers, or in the center of the floor may indicate suspicious activity.*)	Are there signs that preparation was made for a fire? (*Is there an absence or shortage of clothing, furnishings, appliances, food, dishes, personal possessions, items of sentimental value, or pets?*)	Do normal household items appear to be removed or replaced with junk or items of inferior quality? Were major appliances disconnected or unplugged?	Is there obsolete equipment or inventory, fixtures, display cases, equipment, or raw materials?	Are important business records out of their normal places or left where they would be endangered by fire? Are safes, fire-resistant files, etc., open and exposing the contents?

Practical Activity Sheet 17-1
Assess the Origins and Causes of Fires

Name _____ Date _____

Evaluator _____ Overall Competency Rating _____

References	NFPA 1001, Fireground Operations 4-3.4b *Essentials*, pages 621–627
Prerequisites	None
Introduction	One of the most important steps in discovering a fire's cause is to determine its likely site of origin. Examining evidence at that site is likely to reveal the probable source of ignition. Whether a fire is accidental or deliberately set, determining its cause can provide departments with valuable information about the need for equipment, training, operating procedures, public education, and other requirements. This practical activity sheet will help you gain skills to assess fire cause and origin.
Directions	Read each scenario. Use the information provided to determine the fire's likely cause and point of origin. Answer the questions provided with each situation. Write your answers on the blanks. When you feel that you have mastered these skills, schedule an evaluation date with your instructor.
Activity	SCENARIO 1: Upon arrival, you find that a house intensely burning. The west end, including the attached garage, kitchen, and living room, is fully involved with fire spreading to the bedrooms on the east end of the house. Soon after the attack begins, the garage collapses. The fire is extinguished with water. During overhaul, the garage is found to contain the burned hulk of a 1957 Chevrolet, beside which are the remains of a man, later identified as the owner of the house. There are car parts and metal pans around the body. A few feet *(meters)* away is a gas water heater with much of its outer metal skin warped and severely scorched. Along the wall on both sides of the water heater, the fire has completely consumed the gypsum wallboard and most of the wood studs of the house structure.

1. Where is the probable site of origin for this fire?

   _____

2. What is the probable cause of this fire?

   _____

   _____

3. If there is reason to suspect arson in this incident, what is the potential evidence of arson?

_____

_____

_____

_____

4. What action should the department consider as a result of this fire (follow-up investigation, education, training, etc.)

_____

_____

_____

_____

SCENARIO 2: The department has received a 9-1-1 call from an apartment building that has a local alarm system. When you reach the alarm site, a four-story apartment building is on fire. Flames are coming from the open windows of three apartments on the second floor (206, 208, and 210) and one on the third floor (308). While responding, firefighters find that the stairwell doors have been propped open and that the manual pull station on the second floor has been actuated. The fire extinguisher from that site is missing but is later found partially discharged in a stairwell. After the fire is extinguished, the damage is assessed as follows:

- *Apartment 206:* The draperies burned, and the exterior wall in the bedroom is heavily scorched with some charring of the structural members. The interior wall adjoining Apartment 208 is heavily charred. Moderate smoke and water damage has occurred. The occupants state that they were walking their dog and found when they returned that the building was on fire and had been evacuated.

- *Apartment 208:* The draperies burned, and the exterior wall in the bedroom is heavily scorched with heavy charring of the structural members. Damage is concentrated on the wall at the head of the bed. The mattress has a large burn hole in it with the top cloth covering completely consumed by fire. The ceiling is blackened with soot, and there is charring directly over the bed. The carpet is burned in a wide semicircle around the bed. The walls with the adjoining apartments are scorched and charred along the exterior wall. Heavy smoke and water damage has occurred. The occupant states that he was not at home at the time of the fire and returned to the building to find that it had been evacuated.

- *Apartment 210:* The draperies burned, and the exterior wall in the bedroom is heavily scorched with some charring of the structural members. The interior wall adjoining Apartment 208 is heavily charred. Moderate smoke and water damage has occurred. The occupants were not available at the time of the fire but reported later when interviewed that they had been out for the evening.

- *Apartment 308:* The draperies burned, and the exterior wall in the bedroom is moderately scorched. Moderate smoke and water damage has occurred. The occupant states that he was watching television in his living room when he heard the fire alarm. He grabbed some possessions from his bedroom before leaving and noticed flames outside his bedroom window.

1. Where is the probable site of origin for this fire?

   _____

2. What is the probable cause of this fire?

   _____

   _____

3. If there is reason to suspect arson in this incident, what is the potential evidence of arson?

   _____

   _____

   _____

   _____

4. What action should the department consider as a result of this fire (follow-up investigation, education, training, etc.)

   _____

   _____

   _____

   _____

SCENARIO 3: A locally owned photocopy shop has been totally destroyed by fire. The most heavily damaged parts of the building are a closet used to store cleaning materials and copier chemicals, the paper storage area, and the service counter. There is heavy charring of the floor in a line from the storage closet to the paper storage area. The service counter was a wood frame structure with plywood sides, particleboard shelves, and a laminate-covered top. While the response is in progress, the owner of the shop and building arrives at the site. He is particularly concerned about the loss of business records, which were in a box on the service counter in preparation for a meeting he was to have the next day with his accountant. He also asks how long it will be before there is a report available from the fire department so that he can file his insurance claim. During overhaul, firefighters express surprise at how quickly the service counter became involved and was consumed. One remarked, *"I'm surprised that this place was able to make any money as old as some of this equipment is."*

1. Where is the probable site of origin for this fire?

   _____

2. What is the probable cause of this fire?

   _____

   _____

3. If there is reason to suspect arson in this incident, what is the potential evidence of arson?

_____

_____

_____

_____

4. What action should the department consider as a result of this fire (follow-up investigation, education, training, etc.)

_____

_____

_____

_____

SCENARIO 4: A restaurant catches on fire during the night. The first report is a 9-1-1 call from an individual who refuses to identify herself. Upon arrival, firefighters find the roof over the kitchen at the rear of the building heavily involved. Several times during the course of the fire, flames flare up through the roof vents. When the fire is extinguished, an examination of the facility shows heavy charring of the ceiling and roof structure. The filters and exhaust-fan housings over the stoves and grills have collapsed and dropped to the floor and lie on top of the grills. The filters themselves are entirely consumed by fire, and one of the metal mesh screens on a filter assembly has burned through under extremely intense heat. There is little damage to the plaster walls, though there is heavy soot near the ceiling line. The business owner arrives during the fire and tells the incident commander that she is sure that her business was compliant with all regulations.

1. Where is the probable site of origin for this fire?

_____

2. What is the probable cause of this fire?

_____

_____

3. If there is reason to suspect arson in this incident, what is the potential evidence of arson?

_____

_____

_____

_____

4. What action should the department consider as a result of this fire (follow-up investigation, education, training, etc.)

_____

_____

_____

_____

SCENARIO 5: A fire at an abandoned warehouse is well underway when reported in the middle of the night during a January blizzard. The fire is so intense and widespread upon the responders' arrival that they can do little but protect exposures. The structure is weakened to the point that building collapse is likely. A police officer at the scene tells the incident commander that the warehouse has long been used by transients for shelter and that they frequently build fires in the structure on nights such as this to keep warm. A survey of the site before the start of overhaul procedures reveals several areas where unknown persons may have taken shelter with the remnants of mattresses, blankets, cardboard boxes, cooking utensils, and food. Although the entire building is severely damaged, the investigator is particularly interested in the heavy charring of the building's wooden support columns and piles of cardboard and cloth ash and charred lumber at the bases of half the columns. One firefighter suggests that the lumber and cardboard are often used by transients to make lean-tos and similar shelters. During overhaul, an empty gasoline can is found in a collapsed section of the building.

1. Where is the probable site of origin for this fire?

_____

2. What is the probable cause of this fire?

_____

_____

3. If there is reason to suspect arson in this incident, what is the potential evidence of arson?

_____

_____

_____

_____

4. What action should the department consider as a result of this fire (follow-up investigation, education, training, etc.)

_____

_____

_____

_____

Competency Rating Scale

3 — Skilled — "Yes" checked for all 20 criteria; student requires no additional practice.

2 — Moderately skilled — "Yes" checked for at least 3 criteria in each scenario; student may benefit from additional practice.

1 — Unskilled — Fewer than 3 criteria checked "yes" in each scenario; student requires additional practice and reevaluation.

☒ —**Unassigned** — Task is not required or has not been performed.

✔ **Evaluator's Note:** Score the product as indicated below. Use the rating scale above to assign an overall competency rating. Record the overall competency rating on both the student's practical activity sheet and competency profile.

To show competency in this objective, the student must achieve an overall ratinig of at least 2

Criteria	Yes	No
All answers evaluated per suggested answers in the Instructor's Guide.		
SCENARIO 1		
Identified probable site of fire origin	☐	☐
Identified probable cause of fire	☐	☐
Identified potential evidence of arson	☐	☐
Appropriate actions determined	☐	☐
SCENARIO 2		
Identified probable site of fire origin	☐	☐
Identified probable cause of fire	☐	☐
Identified potential evidence of arson	☐	☐
Appropriate actions determined	☐	☐
SCENARIO 3		
Identified probable site of fire origin	☐	☐
Identified probable cause of fire	☐	☐
Identified potential evidence of arson	☐	☐
Appropriate actions determined	☐	☐
SCENARIO 4		
Identified probable site of fire origin	☐	☐
Identified probable cause of fire	☐	☐
Identified potential evidence of arson	☐	☐
Appropriate actions determined	☐	☐
SCENARIO 5		
Identified probable site of fire origin	☐	☐
Identified probable cause of fire	☐	☐
Identified potential evidence of arson	☐	☐
Appropriate actions determined	☐	☐

Job Sheet 17-1
Protect Evidence of Fire Cause and Origin

Name _____ Date _____

Evaluator _____ Overall Competency Rating _____

References	NFPA 1001, Fireground Operations 4-3.4 *Essentials*, page 629
Prerequisites	None
Student's Instructions	To meet evaluation standards, you must perform this job within ____ *[amount of time, if applicable]*; you may have ____ attempts. When you are ready to perform this job, ask your instructor to observe the procedure and complete this form. To show mastery of this job, you must perform all steps to receive an overall competency rating of at least 2.

> **Competency Rating Scale**
>
> 3 — **Skilled** — Meets all evaluation criteria and standards; performs task independently on first attempt; requires no additional practice or training.
>
> 2 — **Moderately skilled** — Meets all evaluation criteria and standards; performs task independently; additional practice is recommended.
>
> 1 — **Unskilled** — Is unable to perform the task; additional training required.
>
> ☒ — **Unassigned** — Job sheet task is not required or has not been performed.
>
> ✔ **Evaluator's Note:** Formulate and inform the candidate of the standards for this task (time allowed and number of attempts). Observe the candidate perform the task, check the step/key point under the appropriate attempt number as accomplished, record total time (if appropriate), and then use the rating scale above to assign an overall competency rating. If the candidate is unable to perform any step of this job, have the candidate review the materials and try again.

Introduction	Arsonists continue to become more sophisticated in their methods, making it harder and harder to detect their crimes. Consequently, gathering evidence is more critical than ever. During all phases of fire fighting, firefighters are in the best position to discover and protect evidence. This job sheet gives you practice in protecting evidence that is likely to be found at a fire site.
Equipment and Personnel	✔**Note:** The need to protect evidence can occur at any point during a response, including the attack, salvage and overhaul, or formal investigation. Equipment and personnel needs should be appropriate to the stage of response. • One or more firefighters dressed appropriately for the situation • Paper and pencil or pen • Camera and film • Cardboard boxes • Plastic sheeting

Job Steps	Key Points	Attempt No. 1 2 3

1. Protect potential evidence.

    1. a. Avoiding touching   ___ ___ ___

        b. Avoiding disturbing   ___ ___ ___

        c. Avoiding tramping   ___ ___ ___

        d. Avoiding the use of excessive water during extinguishment   ___ ___ ___

        e. Leaving evidence in place unless it must be moved to preserve it   ___ ___ ___

2. Preserve evidence as necessary.

    2. a. Covering footprints and tire tracks with cardboard boxes   ___ ___ ___

        b. Covering loose papers and other evidence lightly with plastic sheeting to protect from drafts and water   ___ ___ ___

        c. Moving evidence only as necessary to preserve it   ___ ___ ___

        d. Providing security for the evidence until an investigator is available   ___ ___ ___

3. Move evidence as necessary.

    3. a. Avoiding damage to evidence   ___ ___ ___

        b. Providing security for the evidence until an investigator is available   ___ ___ ___

4. Record information about evidence.

    4. a. Documenting information about location and appearance of evidence if it must be moved or cannot be preserved   ___ ___ ___

        b. Initiating chain of custody record if control of evidence is turned over to anyone else   ___ ___ ___

5. Provide evidence and records to investigator.

    5. Before leaving incident site   ___ ___ ___

Time (Total)   ___ ___ ___

Evaluator's Comments _____

_____

Chapter 17 Review Test

> →**Directions:** This review test covers the Firefighter II material in Chapter 17 of your ***Essentials of Fire Fighting*** text. It may be assigned as a study aid (self-test) or may be administered by your instructor as a pretest or posttest.
>
> When used as a study aid, try to answer the questions without referring to the page numbers in ***Essentials*** or your ***Firefighter II Student Applications*** workbook *(SA)* on which the answers can be found until after you have completed the entire test. Then check your answers against those on the pages provided in parentheses.
>
> When administered by your instructor as a pretest or posttest, read each of the test questions carefully. Choose the best response and then darken the corresponding letter on your answer sheet.
>
> This chapter review test contains 25 multiple-choice questions, each worth 4 points. To pass the test, you must achieve at least 84 of the 100 points possible.

1. What factors contribute to the cause of any fire? *(621)*

    A. The fuel, the source of ignition, and the source of oxygen

    B. The fuel, the source of ignition, and the act or omission that brought the other two factors together

    C. How the fuel was ignited, when it was ignited, and where it was ignited

    D. How the fuel was ignited, where it was ignited, and who was responsible for igniting it

2. Who has the best opportunity to observe evidence of cause and to assist in determination? *(621)*

    A. Fire officer

    C. Fire investigator

    B. Firefighter involved in the attack

    D. Insurance investigator

3. Which fires should a jurisdiction investigate? *(621)*

    A. Fires that result in death

    B. Fires that require more than one alarm

    C. Fires that are suspicious in origin

    D. All fires

4. Firefighter A says that fire fighting efforts may impair the investigation of cause.

    Firefighter B says that firefighters are not able to take actions to protect cause evidence during tactical operations.

    Who is right? *(621)*

    A. Firefighter A

    C. Both A and B

    B. Firefighter B

    D. Neither A nor B

---

5. In most jurisdictions, who has the legal responsibility for determining the cause of a fire? *(622)*

   A. The mayor
   B. The fire chief
   C. The fire marshal
   D. The director of public safety

6. With whom does the responsibility to conduct an interview with a suspected arsonist rest*? *(627)*

   A. State fire marshal
   B. First-arriving firefighter
   C. Police department
   D. Trained investigator

7. To whom should a firefighter or fire officer make statements of accusation, personal opinion, and probable cause? *(627)*

   A. To the property owner
   B. To the press
   C. To the property's insurer
   D. To the fire investigator

8. Why should firefighters never prematurely announce fire cause or make statements in jest at the scene? *(627)*

   A. Fire cause is not a matter for public dissemination.
   B. Such statements may embarrass the fire department and hamper the legal process.
   C. Such statements should be reserved for the privacy of the fire station.
   D. Such statements can only be made by the chief.

9. What is a sufficient reply to any question concerning fire cause? *(627)*

   A. *"The fire is under investigation."*
   B. *"No comment."*
   C. *"It is unlikely that we will be able to determine the cause of this fire."*
   D. *"That information is confidential fire department data not releasable to the public."*

10. How long should the premises be guarded and kept under the control of the fire department?

    A. Until all evidence has been gathered and evaluated by the fire investigator exactly as it appears
    B. For 60 days after completion of the operation
    C. Until after all court proceedings regarding the incident have been completed
    D. Until the property owner insists that the department vacate the premises

11. Firefighter A says that if an investigator is not immediately available, firefighters should leave the premises and ask law enforcement personnel to protect the scene.

    Firefighter B says firefighters should maintain a record of everyone who enters the premises, the time and duration of their visit, and a description of anything that they remove from the site.

    Who is right? *(628)*

    A. Firefighter A
    B. Firefighter B
    C. Both A and B
    D. Neither A nor B

12. Why should all evidence be marked, tagged, and photographed before the last firefighter leaves the scene? *(628)*

A. Because that is a normally assigned duty of the last department person

B. Because in many instances a search warrant or written consent to search will be needed for further visits to the premises

C. So that such activities do not interfere with tactical operations, they should be done last

D. To prevent damage to the delicate measuring instruments used in the collection of evidence

13. When does a department's right to deny access to an incident site end? *(628)*

A. As soon as the property owner asks the department to leave

B. When the last department representative leaves the property

C. When the department formally releases the property to the owner

D. As soon as the investigation is complete

14. Firefighter A says that firefighters should secure the site and maintain security.

Firefighter B says that departments may use law enforcement personnel to secure sites.

Who is right? *(628)*

A. Firefighter A   C. Both A and B

B. Firefighter B   D. Neither A nor B

15. When the scene is secured, who should be allowed to enter the premises? *(628)*

A. Only department personnel and members of the news media

B. No one unless accompanied by an authorized individual within local laws as applicable to property owners

C. Only law enforcement officers

D. No one except the fire investigators

16. When a person is authorized entry to the premises, what information should be recorded in the entry log? *(628)*

A. Person's name, social security number, and description of any items taken from the scene

B. Person's name, times of entry and departure, and description of any items taken from the scene

C. Person's name, times of entry and departure, and vehicle license number

D. Person's name, physical description, and social security number

17. On what legal case was the decision to require a search warrant on return to premises based? *(628)*

A. Minnesota *vs.* Taylor (1988)

B. Mississippi *vs.* Tyson (1987)

C. Wisconsin *vs.* Towson (1977)

D. Michigan *vs.* Tyler (1978)

18. What is the likely legal impact of reentering the premises without the owner's written consent or a search warrant? *(629)*

    A. Arrest of the fire chief

    B. Dissolution of the department

    C. Making prosecution or conviction of the alleged arsonist impossible

    D. The need to provide a chain of custody report to the court in order to admit evidence

19. Because of the variations in local legislation with regard to department access to property and authorization to retain possession of property, jurisdictions should ___. *(628, 629)*

    A. Develop SOPs around the legislation and legal opinions that affect the jurisdiction.

    B. Prolong overhaul actions as long as possible.

    C. Obtain court orders to retain possession of property for 30 days following an incident.

    D. Obtain court orders to retain possession of property for 60 days following an incident.

20. What is the preferred way for a firefighter to deal with potential evidence that is found on site? *(629)*

    A. To immediately recover it and place it in an area of safe storage

    B. To immediately cease salvage operations and notify the IC of the evidence so that Command can determine what to do

    C. To cordon off the area during daytime operations or to mark it with a flare during nighttime operations

    D. To leave it untouched and undisturbed and to provide security for the area

21. How large an area should be cordoned off and controlled at the incident site during the investigation? *(628)*

    A. 100 yards *(90 m)* in all directions from the point of origin

    B. 100 yards *(90 m)* in all directions beyond the farthest limits of the area affected by the incident

    C. At the outside perimeter of the property as legally defined in the deed

    D. There are no specifically defined boundaries for the cordon.

22. Under what circumstances should a firefighter gather and handle physical evidence? *(629)*

    A. Whenever such evidence is discovered

    B. Only if it is absolutely necessary to preserve the evidence

    C. Whenever there will be more than a 12-hour delay in the arrival of the investigator

    D. Only if absolutely necessary to allow thorough overhaul and salvage

23. What must a firefighter do when he or she handles or procures evidence? *(629)*

    A. Undergo decontamination.

    B. Accurately document all actions taken as soon as possible.

    C. Testify in all related court proceedings.

    D. Notify the fire investigator immediately.

24. What changes in the evidence should be allowed? *(629)*

    A. Those absolutely necessary in the extinguishment of the fire

    B. Those absolutely necessary to improve the quality of photographs that can be obtained

    C. Those absolutely necessary to remove evidence from toxic atmospheres

    D. None

25. What is meant by chain of custody? *(629)*

    A. The line of authority in a jurisdiction

    B. The sequence of persons who have had access to evidence

    C. The process for detaining, interrogating, and arresting or releasing an arson suspect

    D. The packaging of evidence for preservation

REVIEW TEST ANSWER SHEET

	A	B	C	D
1.	○	○	○	○
2.	○	○	○	○
3.	○	○	○	○
4.	○	○	○	○
5.	○	○	○	○
6.	○	○	○	○
7.	○	○	○	○
8.	○	○	○	○
9.	○	○	○	○
10.	○	○	○	○
11.	○	○	○	○
12.	○	○	○	○
13.	○	○	○	○
14.	○	○	○	○
15.	○	○	○	○
16.	○	○	○	○
17.	○	○	○	○
18.	○	○	○	○
19.	○	○	○	○
20.	○	○	○	○
21.	○	○	○	○
22.	○	○	○	○
23.	○	○	○	○
24.	○	○	○	○
25.	○	○	○	○
26.	○	○	○	○
27.	○	○	○	○
28.	○	○	○	○
29.	○	○	○	○
30.	○	○	○	○
31.	○	○	○	○
32.	○	○	○	○
33.	○	○	○	○

	A	B	C	D
34.	○	○	○	○
35.	○	○	○	○
36.	○	○	○	○
37.	○	○	○	○
38.	○	○	○	○
39.	○	○	○	○
40.	○	○	○	○
41.	○	○	○	○
42.	○	○	○	○
43.	○	○	○	○
44.	○	○	○	○
45.	○	○	○	○
46.	○	○	○	○
47.	○	○	○	○
48.	○	○	○	○
49.	○	○	○	○
50.	○	○	○	○
51.	○	○	○	○
52.	○	○	○	○
53.	○	○	○	○
54.	○	○	○	○
55.	○	○	○	○
56.	○	○	○	○
57.	○	○	○	○
58.	○	○	○	○
59.	○	○	○	○
60.	○	○	○	○
61.	○	○	○	○
62.	○	○	○	○
63.	○	○	○	○
64.	○	○	○	○
65.	○	○	○	○
66.	○	○	○	○
67.	○	○	○	○

	A	B	C	D
68.	○	○	○	○
69.	○	○	○	○
70.	○	○	○	○
71.	○	○	○	○
72.	○	○	○	○
73.	○	○	○	○
74.	○	○	○	○
75.	○	○	○	○
76.	○	○	○	○
77.	○	○	○	○
78.	○	○	○	○
79.	○	○	○	○
80.	○	○	○	○
81.	○	○	○	○
82.	○	○	○	○
83.	○	○	○	○
84.	○	○	○	○
85.	○	○	○	○
86.	○	○	○	○
87.	○	○	○	○
88.	○	○	○	○
89.	○	○	○	○
90.	○	○	○	○
91.	○	○	○	○
92.	○	○	○	○
93.	○	○	○	○
94.	○	○	○	○
95.	○	○	○	○
96.	○	○	○	○
97.	○	○	○	○
98.	○	○	○	○
99.	○	○	○	○
100.	○	○	○	○

Name _____

Date _____

Score _____

Chapter 17 Competency Profile

Student Name _____ Soc. Sec. No. _____
                    Last   First   Middle

Fire Department _____

Address _____

Phone _____

Home Address _____

Phone _____

Date of Enrollment _____ - _____ - _____   Total Class Hours _____

Date of Withdrawal _____ - _____ - _____   Total Hours Absent _____

Date of Completion _____ - _____ - _____

Instructor's Name _____   Session Dates _____

Instructor's Directions

1. Check the candidate's competency rating (3, 2, 1, ☒) for each performance test task and psychomotor lesson objective (practical activity and job sheets) listed below.

2. List any additional performance tasks or psychomotor objectives (job sheets or practical activity sheets) under "Other," and check the candidate's competency rating.

3. Record the candidate's cognitive scores (written lesson tests and *administered* chapter review tests) in the spaces provided.

Level				Psychomotor Competencies
3	2	1	☒	

Practical Activity Sheets

☐ ☐ ☐ ☐ PAS 17-1 — Assess the Origins and Causes of Fires

☐ ☐ ☐ ☐ Other _____

☐ ☐ ☐ ☐ _____

Job Sheets

☐ ☐ ☐ ☐ JS 17-1 — Protect Evidence of Fire Cause and Origin

☐ ☐ ☐ ☐ Other _____

☐ ☐ ☐ ☐ _____

Chapter 17 Performance Test

☐ ☐ ☐ ☐ Task 1 — Determine origin and cause of fire.

☐ ☐ ☐ ☐ Task 2 — Protect evidence.

Level

3   2   1   ☒

☐   ☐   ☐   ☐

☐   ☐   ☐   ☐

Psychomotor Competencies

Other _____

_____

Points Achieved	Points Needed/ Total	Cognitive Competencies

Written Test

_____ 2/4   1. List responsibilities of a fire investigator.

_____ 3/4   2. Select facts about conduct and statements at the scene.

_____ 8/9   3. Select facts about securing the scene and legal considerations.

_____ 4/5   4. Select facts about protecting and preserving evidence.

5. Evaluated on Job Sheet 17-1

6. Evaluated on Practical Activity Sheet 17-1

Review Test

_____   Chapter 17 Review Test

Instructor's Signature _____ Date _____

Student's Signature _____ Date _____

STUDENT APPLICATIONS

FOURTH EDITION

ESSENTIALS OF FIRE FIGHTING

LESSON

18

RADIO COMMUNICATIONS & INCIDENT REPORTS

FIREFIGHTER II

FIRE PROTECTION PUBLICATIONS
OKLAHOMA STATE UNIVERSITY

Study Objectives

LESSON OBJECTIVE

After completing this lesson, you will be able to complete a basic incident report and communicate the need for team assistance.

ENABLING OBJECTIVES

After reading Chapter 18 of *Essentials*, pages 649 and 651, and completing related activities, you will be able to —

1. Select facts about making calls for additional response.

2. List information that should be included in incident reports.

3. **Identify appropriate incident report codes. *(Practical Activity Sheet 18-1)***

4. **Proofread incident reports. *(Practical Activity Sheet 18-2)***

5. **Create incident reports using department equipment. *(Practical Activity Sheet 18-3)***

Study Sheet

Introduction | This study sheet is intended to help you learn the Firefighter II material in Chapter 18 of *Essentials of Fire Fighting*, Fourth Edition. You may use it for self-study, or you may use it to review material that will be covered in the lesson and chapter review tests. The numbers in parentheses are the pages in *Essentials* on which the answers or terms can be found.

Chapter Vocabulary | Be sure you know the chapter-related meanings of the following terms and abbreviations. Use a dictionary or the glossary in *Fire Service Orientation and Terminology* if you cannot determine the meaning of the term from its context.

- Additional alarms *(649)*
- Database *(651)*
- Incident report *(651)*
- NFIRS *(651)*
- PC *(651)*
- USFA *(651)*

Study Questions & Activities

1. Normally, who may order multiple alarms or additional response to a fire? *(649)*

   _____

   _____

2. What are the procedures in your area for requesting additional alarms? *(Local protocol)*

   _____

   _____

   _____

3. List the information that should be included in incident reports. *(651)*

   a. _____

   _____

   b. _____

   _____

   c. _____

   _____

   d. _____

   _____

e. _____

_____

f. _____

_____

g. _____

_____

h. _____

_____

i. _____

_____

j. _____

_____

k. _____

_____

l. _____

_____

m. _____

_____

Practical Activity Sheet 18-1
Identify Appropriate Incident Report Codes

Name _____ Date _____

Evaluator _____ Overall Competency Rating _____

References	NFPA 1001, Fire Department Communications 4-2.1a *Essentials,* page 651
Prerequisites	None
Introduction	To provide some consistency to statistical databases, departments have adopted standardized incident reporting systems for the information gained through emergency responses. Codes can be used to indicate the type of emergency, cause, attack method, fuels, complications, injuries, and many other aspects of an incident. Firefighters responsible for generating incident reports must know how to use the codes associated with them.
Directions	Read each situation described and then answer the questions that accompany them. Your answers should reflect the incident reporting system used by your department. Write your answers on the blanks. ✔Note: Refer to *Essentials*, Figure 18.29, page 650, for additional information about the department and properties.
Activity	SITUATION 1: On March 19, at 0436, an automatic alarm is received from Heimbach Brothers Hardware Store, 301 Main Street. A one-alarm response is made. However, upon arrival at 0440, the commander requests a second alarm. The fire is extinguished after an aggressive attack with hoselines. The back half of the building is destroyed with heavy smoke damage and slight water damage to the front half of the building. The residence north of the hardware store suffers slight water damage from efforts to protect it as an exposure. The origin of the fire is a short in the main circuit breaker box, aggravated by the storage of paper and flammable chemicals in an adjoining closet. During the response, two firefighters are injured but not hospitalized when a section of roof collapses during ventilation. One firefighter suffers a sprained ankle and cuts and bruises, while the other sustains a dislocated finger and bruises during the collapse. The collapse appears to have resulted from the failure of fire-weakened wooden trusses.

1. Based on your jurisdiction's incident reporting system, identify the report codes for each of the following:

   a. Type of emergency _____

   b. Type of structure or site _____

   c. Cause of emergency _____

   d. Action that was taken _____

 e. Property use _____

 f. Types of injuries/fatalities_____

 g. Cause of injuries/fatalities _____

 h. Method used to extinguish fire _____

 i. Type of apparatus that responded _____

 j. Method of receiving call_____

SITUATION 2: At 1426, August 19, a call is received from the Valley Shoe Store reporting a fire at the Funtimes Novelty Shop, 313 Main Street. An alarm is issued. The first unit arrives at 1431 to find the building filled with smoke and several people standing around on the sidewalk. Upon entering the building, firefighters find that the source of the smoke is a number of smoke bombs that have been activated. There is a small fire that resulted from smoke bomb fuses coming into contact with paper products, but it is quickly extinguished with a fire extinguisher. The smoke bombs appear to have been deliberately ignited. Two customers are treated for cuts received when the door glass is broken during the evacuation. One of the victims is taken to the hospital emergency room and released after receiving stitches.

1. Based on your jurisdiction's incident reporting system, identify the report codes for each of the following:

 a. Type of emergency _____

 b. Type of structure or site _____

 c. Cause of emergency _____

 d. Action that was taken _____

 e. Property use _____

 f. Types of injuries/fatalities _____

 g. Cause of injuries/fatalities _____

 h. Method used to extinguish fire _____

 i. Type of apparatus that responded _____

 j. Method of receiving call _____

Competency Rating Scale

3 — Skilled — "Yes" checked for all criteria; student requires no additional practice.

2 — Moderately skilled — "Yes" checked for at least 8 criteria in each situation; student may benefit from additional practice.

1 — Unskilled — Fewer than 8 criteria for each situation checked "yes"; student requires additional practice and reevaluation.

☒ — Unassigned — Task is not required or has not been performed.

✔ **Evaluator's Note:** Score the product as indicated below. Use the rating scale above to assign an overall competency rating. Record the overall competency rating on both the student's practical activity sheet and competency profile.

To show competency in this objective, the sutdent must achieve an overall rating of at least 2.

Responses	Correct	Incorrect
Candidate responded with correct code for item listed, per local protocol.		
SITUATION 1		
a. Type of emergency	☐	☐
b. Type of structure or site	☐	☐
c. Cause of emergency	☐	☐
d. Action that was taken	☐	☐
e. Property use	☐	☐
f. Types of injuries/fatalities	☐	☐
g. Cause of injuries/fatalities	☐	☐
h. Method used to extinguish fire	☐	☐
i. Type of apparatus that responded	☐	☐
j. Method of receiving call	☐	☐
SITUATION 2		
a. Type of emergency	☐	☐
b. Type of structure or site	☐	☐
c. Cause of emergency	☐	☐
d. Action that was taken	☐	☐
e. Property use	☐	☐
f. Types of injuries/fatalities	☐	☐
g. Cause of injuries/fatalities	☐	☐
h. Method used to extinguish fire	☐	☐
i. Type of apparatus that responded	☐	☐
j. Method of receiving call	☐	☐

Practical Activity Sheet 18-2
Proofread Incident Reports

Name _____ Date _____

Evaluator _____ Overall Competency Rating _____

References	NFPA 1001, Fire Department Communications 4-2.1a *Essentials*, page 651
Prerequisites	None
Introduction	The importance of proofing incident reports is twofold. First, mistakes in reports — including typographical errors — can result in inaccurate reports. For example, transposing two numbers could affect the date of the incident, the companies that responded, the number of injuries, the address, the estimated cost of damages, or other values in the report. Secondly, incident reports are used in court proceedings, insurance claims, and other activities that provide public exposure to the department. Poorly written and edited reports can be a tremendous embarrassment to the department.
Directions	Read the comments section of the incident report shown below and on the next page. Rewrite the report in the space provided. Correct defects in the original text, including grammatical errors, spelling errors, inaccurate information, passages that are not clear, inappropriate comments, etc. ✔**Note:** The sentences in the report have been numbered for convenience in discussing the report in class. Use Figure 18.29, page 650, in *Essentials* as an additional resource about the property. The following dispatch log entry is accurate and can be used to verify some of the information in the incident report: **Date:** 6/6 **Time:** 1500 **Action:** Dispatched E-22, S-61, M-369 to report of vehicle/structure fire with probable injury. Chiaro's Pizza, 338 Main. Reported to 9-1-1 by G. Nedry, Assistant Manager at Chiaro's.
Activity	**INCIDENT REPORT COMMENTS:** (1) Nothing noteworthy about the incident. (2) A hoseline was placed on the front of Chirao's to provide protection to the vehicle extrication team and extingish the fire. (3) We had it down in less than 10 minutes. (4) We had to take maybe a dozen people out the back door since the front was all messed up by the wreck. (5) The driver was unconscious and had to be pulled out. (6) One of the cops told me he was drunk as a skunk. (7) Most of the injuries were from flying glass and debree. (8) They were all treated by the 369 guys, except two people who were acting kind of funny. (9) Who were taken to the hospital to see if they were in shock or had concusions. (10) One engine guy had to set down for a while because he got overheated. (11) It was one hot sucker out there today. (12) We were lucky though because those geeks over at at Street Maintnance had a detour on 3rd that nobody told us about and we had to go around on Preston.

(13) That delay could of let the situation get out of hand but we did okay. (14) After the fire was out, we pulled the car away from the building using the winch on the front of 22 even though it's not strictly by the book. (15) Then the wrecker boys hauled it off so we could do salvage. (16) It'll probably be a couple of months before Chiaro's is back in business.

Rewrite: _____

_____

_____

_____

_____

_____

_____

_____

_____

_____

_____

_____

_____

_____

_____

_____

_____

_____

_____

_____

_____

_____

_____

_____

Competency Rating Scale

3 — Skilled — All sentences corrected appropriately per suggested answers in *Instructor's Guide*; student requires no additional practice.

2 — Moderately skilled — At least 12 of the 16 sentences corrected appropriately per suggested answers in *Instructor's Guide*; student may benefit from additional practice.

1 — Unskilled — Fewer than 12 sentences corrected appropriately per suggested answers in *Instructor's Guide*; student requires additional practice and reevaluation.

☒ — **Unassigned** — Task is not required or has not been performed.

✔ **Evaluator's Note:** Score the product as indicated below. Use the rating scale above to assign an overall competency rating. Record the overall competency rating on both the student's practical activity sheet and competency profile.

To show competency in this objective, the student must achieve an overall rating of at least 2.

Revisions	Correct	Incorrect
*Candidate's sentence revisions appropriate according to answers in **Instructor's Guide** materials.*		
Sentence 1	☐	☐
Sentence 2	☐	☐
Sentence 3	☐	☐
Sentence 4	☐	☐
Sentence 5	☐	☐
Sentence 6	☐	☐
Sentence 7	☐	☐
Sentence 8	☐	☐
Sentence 9	☐	☐
Sentence 10	☐	☐
Sentence 11	☐	☐
Sentence 12	☐	☐
Sentence 13	☐	☐
Sentence 14	☐	☐
Sentence 15	☐	☐
Sentence 16	☐	☐

Practical Activity Sheet 18-3
Create Incident Reports Using Department Equipment

Name _____ Date _____

Evaluator _____ Overall Competency Rating _____

References	NFPA 1001, Fire Department Communications 4-2.1a *Essentials,* page 651
Prerequisites	None
Introduction	Personnel responsible for generating incident reports must be able to use the department's available equipment. In this assignment, you will use the forms, equipment, and resources available in your jurisdiction to develop an incident report.
Directions	Read each of the following scenarios. Choose one scenario and, based on the facts presented and the *resources available in your jurisdiction,* develop an incident report. You will have to create some information based on anticipated conditions and actions. When you feel that you have mastered these skills, schedule an evaluation date with your instructor.
Activity	SCENARIO 1 On Monday, at 1600 hours, a full first-alarm assignment is dispatched to the Valley Shoe Store, 4412 Main, on a report of smoke in the building. On arrival, Engine 65-22 finds light smoke in the sales area due to a defective ballast in a fluorescent light fixture. A small crowd of shoppers and onlookers has gathered outside the store. SCENARIO 2 At 1500 hours on a summer Friday, Engine 65-22, Squad 65-61, and Medic 369 are dispatched to a report of a vehicle accident/fire at 3rd and Main. On arrival, they find that an automobile has crashed through a wall of Chiaro's Pizza & Apartments at 4414 Main. The driver of the car and several patrons of the pizza shop have been injured; some are walking about in shock or dazed. The vehicle is burning and has ignited a display area in the gift shop. A large crowd of curious shoppers and onlookers has gathered outside the apartments and across the street at the row house.

Competency Rating Scale

3 — Skilled — "Yes" checked for all criteria; student requires no additional practice.

2 — Moderately skilled — "Yes" checked for at least 4 of the 5 criteria; student may benefit from additional practice.

1 — Unskilled — Fewer than 4 criteria checked "yes"; student requires additional practice and reevaluation.

☒ — Unassigned — Task is not required or has not been performed.

✔ Evaluator's Note: Score the product as indicated below. Use the rating scale above to assign an overall competency rating. Record the overall competency rating on both the student's practical activity sheet and competency profile.

To show competency in this objective, the student must achieve an overall competency rating of at least 2.

Criteria	Yes	No
1. Report was complete.	☐	☐
2. Report was accurate.	☐	☐
3. Report complied with SOPs and other local protocol.	☐	☐
4. Related actions (such as updating databases) were performed correctly.	☐	☐
5. Candidate made good and efficient use of equipment.	☐	☐

Chapter 18 Review Test

→ **Directions:** This review test covers the Firefighter II material in Chapter 18 of your *Essentials of Fire Fighting* text. It may be assigned as a study aid (self-test) or may be administered by your instructor as a pretest or posttest.

When used as a study aid, try to answer the questions without referring to the page numbers in *Essentials* or your *Firefighter II Student Applications* workbook *(SA)* on which the answers can be found until after you have completed the entire test. Then check your answers against those on the pages provided in parentheses.

When administered by your instructor as a pretest or posttest, read each of the test questions carefully. Choose the best response and then darken the corresponding letter on your answer sheet.

This chapter review test contains 15 multiple-choice questions, each worth 7 points. To pass the test, you must achieve at least 91 of the 105 points possible.

1. All firefighters should know the local procedures for requesting ___. *(649)*

   A. Transfer to another department   C. All-clear signals

   B. Incident termination   D. Additional alarms

2. Firefighter A says that firefighters need ***not*** know the number of units that respond to alarms.

   Firefighter B says that firefighters need ***not*** know the types of units that respond to alarms.

   Who is right? *(649)*

   A. Firefighter A   C. Neither A nor B

   B. Firefighter B   D. Both A and B

3. Requests for additional assistance should be ordered by ___. *(649)*

   A. The incident commander   C. The command post driver

   B. The operations officer   D. The telecommunicator

4. Each response team should have a supervisor who is in constant contact with the team and who can ___. *(649)*

   A. Follow local IMS and SOPs for communication with the IC and telecommunications center

   B. Terminate the incident on his or her own authority

   C. Sound the evacuation alarm if required

   D. Evaluate the team's compliance with local IMS and SOPs

5. What is one way to reduce the load on the telecommunications center when multiple alarms are given for a single fire? *(649)*

   A. Route communications through a second station's telecommunications center.

   B. Use a mobile, radio-equipped, command vehicle.

   C. Use the local civil defense channel.

   D. Route communications through a local law enforcement agency.

6. When must an incident report be completed? *(651)*

   A. Only when a fire unit responds to an incident that is not a false alarm

   B. Only when a fire unit responds to an incident involving more than $100 damage

   C. Whenever a fire unit responds to an incident

   D. Only when a fire unit responds to an incident that is not successfully concluded before the unit's arrival at the scene

7. Standards for incident reports are covered in ___. *(651)*

   A. NFPA 290   C. NFPA 209

   B. NFPA 920   D. NFPA 902

8. Firefighter A says that incident reports do *not* need to include information about how the emergency was reported.

   Firefighter B says that incident reports should include information about how and where the fire started.

   Who is right? *(651)*

   A. Firefighter A   C. Both A and B

   B. Firefighter B   D. Neither A nor B

9. Which of the following is *not* necessarily included in an incident report? *(651)*

   A. Name(s) of owner(s)

   B. Social Security number(s) of occupants

   C. Addresses of occupants

   D. Property-use information

10. Firefighter A says that incident reports should include information about the estimated cost of damage.

    Firefighter B says that incident reports do *not* need to include information about the number of personnel and type of apparatus that responded.

    Who is right? *(651)*

    A. Firefighter A

    B. Firefighter B

    C. Both A and B

    D. Neither A nor B

11. Who usually writes a narrative of the incident for an incident report? *(651)*

    A. The fire chief

    B. The officer in charge

    C. The public information officer

    D. Each eyewitness

12. Where do most fire departments enter the data from their incident reports? *(651)*

    A. Databases at the local and county level

    B. Databases at the state and national level

    C. Databases at the county and regional level

    D. Databases at the regional and national level

13. Firefighter A says that the information in national databases **cannot** be used to evaluate the needs of individual departments.

    Firefighter B says that the information from departmental databases should **not** be used to justify budget requests.

    Who is right? *(651)*

    A. Firefighter A

    B. Firefighter B

    C. Both A and B

    D. Neither A nor B

14. What is the National Fire Incident Reporting System (NFIRS)? *(651)*

    A. A word processing program that allows easy input of incident information

    B. A national telephone hotline for reporting incident information

    C. An alarm system used for nationwide notification in the event of a national disaster

    D. An Internet-based database that records incident information from the majority of states in the U.S.

15. What agency created the NFIRS? *(651)*

    A. The United States Fire Administration

    B. The National Fire Protection Association

    C. The Occupational Safety and Health Administration

    D. Underwriters Laboratories

REVIEW TEST ANSWER SHEET

	A	B	C	D
1.	○	○	○	○
2.	○	○	○	○
3.	○	○	○	○
4.	○	○	○	○
5.	○	○	○	○
6.	○	○	○	○
7.	○	○	○	○
8.	○	○	○	○
9.	○	○	○	○
10.	○	○	○	○
11.	○	○	○	○
12.	○	○	○	○
13.	○	○	○	○
14.	○	○	○	○
15.	○	○	○	○
16.	○	○	○	○
17.	○	○	○	○
18.	○	○	○	○
19.	○	○	○	○
20.	○	○	○	○
21.	○	○	○	○
22.	○	○	○	○
23.	○	○	○	○
24.	○	○	○	○
25.	○	○	○	○
26.	○	○	○	○
27.	○	○	○	○
28.	○	○	○	○
29.	○	○	○	○
30.	○	○	○	○
31.	○	○	○	○
32.	○	○	○	○
33.	○	○	○	○

	A	B	C	D
34.	○	○	○	○
35.	○	○	○	○
36.	○	○	○	○
37.	○	○	○	○
38.	○	○	○	○
39.	○	○	○	○
40.	○	○	○	○
41.	○	○	○	○
42.	○	○	○	○
43.	○	○	○	○
44.	○	○	○	○
45.	○	○	○	○
46.	○	○	○	○
47.	○	○	○	○
48.	○	○	○	○
49.	○	○	○	○
50.	○	○	○	○
51.	○	○	○	○
52.	○	○	○	○
53.	○	○	○	○
54.	○	○	○	○
55.	○	○	○	○
56.	○	○	○	○
57.	○	○	○	○
58.	○	○	○	○
59.	○	○	○	○
60.	○	○	○	○
61.	○	○	○	○
62.	○	○	○	○
63.	○	○	○	○
64.	○	○	○	○
65.	○	○	○	○
66.	○	○	○	○
67.	○	○	○	○

	A	B	C	D
68.	○	○	○	○
69.	○	○	○	○
70.	○	○	○	○
71.	○	○	○	○
72.	○	○	○	○
73.	○	○	○	○
74.	○	○	○	○
75.	○	○	○	○
76.	○	○	○	○
77.	○	○	○	○
78.	○	○	○	○
79.	○	○	○	○
80.	○	○	○	○
81.	○	○	○	○
82.	○	○	○	○
83.	○	○	○	○
84.	○	○	○	○
85.	○	○	○	○
86.	○	○	○	○
87.	○	○	○	○
88.	○	○	○	○
89.	○	○	○	○
90.	○	○	○	○
91.	○	○	○	○
92.	○	○	○	○
93.	○	○	○	○
94.	○	○	○	○
95.	○	○	○	○
96.	○	○	○	○
97.	○	○	○	○
98.	○	○	○	○
99.	○	○	○	○
100.	○	○	○	○

Name _____

Date _____

Score _____

Chapter 18 Competency Profile

Student Name _____ Soc. Sec. No. _____
                Last   First   Middle

Fire Department_____

Address _____

Phone _____

Home Address _____

Phone _____

Date of Enrollment _____ - _____ - _____   Total Class Hours _____

Date of Withdrawal _____ - _____ - _____   Total Hours Absent_____

Date of Completion _____ - _____ - _____

Instructor's Name _____   Session Dates_____

Instructor's Directions

1. Check the student's competency rating (3, 2, 1, ☒) for each performance test task and psychomotor lesson objective (practical activity and job sheets) listed below.

2. List any additional performance tasks or psychomotor objectives (job sheets or practical activity sheets) under "Other," and check the candidate's competency rating.

3. Record the candidate's cognitive scores (written lesson tests and *administered* chapter review tests) in the spaces provided.

Level				Psychomotor Competencies
3	2	1	☒	

Practical Activity Sheets

3	2	1	☒	
☐	☐	☐	☐	PAS 18-1 — Identify Appropriate Incident Report Codes
☐	☐	☐	☐	PAS 18-2 — Proofread Incident Reports
☐	☐	☐	☐	PAS 18-3 — Create Incident Reports Using Department Equipment
☐	☐	☐	☐	Other _____
☐	☐	☐	☐	_____

Chapter 18 Performance Test

3	2	1	☒	
☐	☐	☐	☐	Task 1 — Complete a basic incident report.
☐	☐	☐	☐	Task 2 — Communicate the need for team assistance.
☐	☐	☐	☐	Other _____
☐	☐	☐	☐	_____

Points Achieved	Points Needed/ Total	Cognitive Competencies

Written Test

_____ 4/5   1. Select facts about making calls for additional response.

_____ 8/8   2. List information that should be included in incident reports.

           3. Evaluated on Practical Activity Sheet 18-1

           4. Evaluated on Practical Activity Sheet 18-2

           5. Evaluated on Practical Activity Sheet 18-3

Review Test

_____ Chapter 18 Review Test

Instructor's Signature _____ **Date** _____

Student's Signature _____ **Date** _____

STUDENT APPLICATIONS

FOURTH EDITION

ESSENTIALS OF FIRE FIGHTING

LESSON
19

PRE-INCIDENT SURVEY

FIREFIGHTER II

FIRE PROTECTION PUBLICATIONS
OKLAHOMA STATE UNIVERSITY

Study Objectives

LESSON OBJECTIVE

After completing this lesson, you will be able to conduct a pre-incident survey, working as a member of a team.

ENABLING OBJECTIVES

After reading Chapter 19 of *Essentials,* pages 655 through 666, and completing related activities, you will be able to —

1. Provide examples of personal traits and skills required of personnel who conduct fire safety surveys.

2. Provide examples of the type of equipment required to conduct fire safety surveys.

3. List goals of pre-incident surveys.

4. Provide examples of the types of information that a pre-incident survey can provide.

5. Match standard map symbols to their correct meanings.

6. **Make field sketch and report drawings. (*Practical Activity Sheet 19-1*)**

7. List objectives of the exit interview during a pre-incident survey.

8. **Perform a pre-incident survey and complete related documentation. (*Job Sheet 19-1*)**

Study Sheet

Introduction

This study sheet is intended to help you learn the Firefighter II material in Chapter 19 of *Essentials of Fire Fighting,* Fourth Edition. You may use it for self-study, or you may use it to review material that will be covered in the lesson and chapter review tests. The numbers in parentheses are the pages in *Essentials* on which the answers or terms can be found.

Chapter Vocabulary

Be sure you know the chapter-related meanings of the following terms and abbreviations. Use a dictionary or the glossary in *Fire Service Orientation and Terminology* if you cannot determine the meaning of the term from its context.

- Elevation sketch *(665, 666)*
- Floor plan sketch *(664)*
- Pitot tube *(661)*
- Plot plan *(664)*
- Pre-incident survey *(659)*
- Target hazard *(659)*
- Water flow tests *(661)*

Study Questions & Activities

1. Name five interpersonal skills firefighters should have to make pre-incident surveys. *(660)*

   a. _____
   _____

   b. _____
   _____

   c. _____
   _____

   d. _____
   _____

   e. _____
   _____

2. Name five technical skills firefightersshould have to make pre-incident surveys? *(660)*

   a. _____
   _____

   b. _____
   _____

   c. _____
   _____

d. _____

_____

e. _____

_____

3. Why is a neat, well-groomed appearance important for survey team members? *(660)*

_____

_____

_____

_____

_____

4. What are some factors that will help a firefighter's survey skills improve? *(660)*

_____

_____

_____

_____

5. List the equipment needed by a firefighter at the following locations to adequately perform a survey. *(660, 661)*

a. Survey site _____

_____

_____

_____

b. Fire station _____

_____

_____

6. What five goals does the information from pre-incident surveys help firefighters achieve? *(662)*

a. _____

_____

b. _____

_____

c. _____

_____

d. _____

_____

e. _____

_____

7. What two contacts with the owner/occupant should occur at both the beginning and during the survey? *(660)*

    a. Beginning of survey _____

        _____

    b. During survey _____

        _____

8. What five observations and checks should be made regarding the exterior of an occupancy? *(662, 663)*

    a. _____

        _____

    b. _____

        _____

    c. _____

        _____

    d. _____

        _____

    e. _____

        _____

9. Should survey teams start at the top or the bottom of a building when conducting the survey? Why? *(663)*

    _____

    _____

    _____

10. Why should a firefighter take sufficient time to make notes and sketches of all important features during a survey? *(663)*

    _____

    _____

    _____

11. Briefly describe the following and tell how they are used. *(664–666)*

    a. Plot plan _____

        _____

        _____

    b. Freehand sketch _____

        _____

        _____

    c. Floor plan _____

        _____

        _____

d. Elevation plan _____

_____

_____

12. How are photographs used in a survey report? *(666)*

_____

_____

13. Why is the exit interview with the owner or person with authority important after a survey? *(666)*

_____

_____

_____

Information Sheet 19-1
Sample Field Sketch

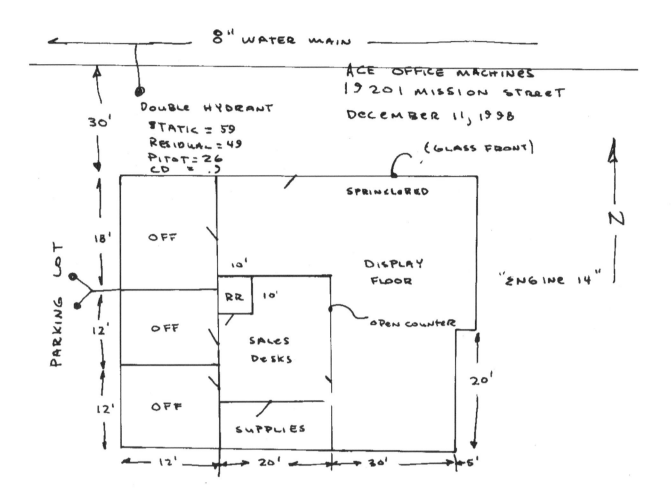

Information Sheet 19-2
Sample Report Drawing

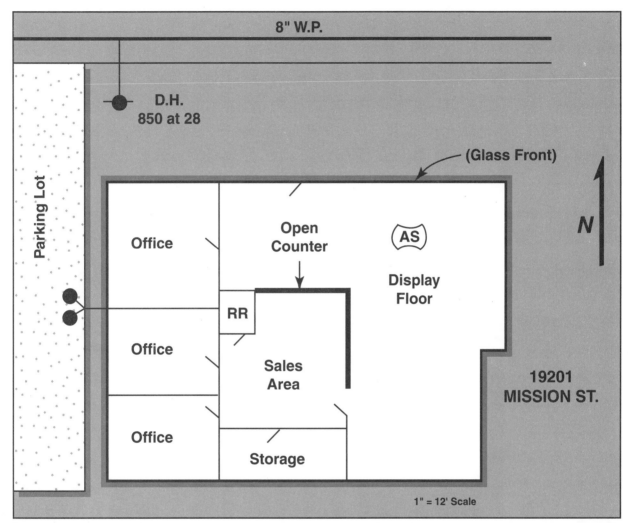

The report drawing should be free of errors and should be drawn tro scale. Details and standardized map symbols are represented. A symbol ledger should be included on or attached to the drawing. The drawing above is not reproduced to scale.

LEGEND

- Two-Way Fire Department Connection
- Public Hydrant, Two Hose Connections
- Concrete
- (AS) Partially Sprinklered Space

Information Sheet 19-3
Checklist for Preparing Field Sketches and Report Drawings

Make sure that all items are included by checking them off when placed on your sketch or report drawing.

☐ North point and scale

☐ Street names

☐ Street numbers

☐ Exposure street numbers, construction, and use

☐ Distances to exposures

☐ Street water mains

☐ Building designation (occupancy numbers and/or letters)

☐ Exterior building dimensions

☐ Windows, doors, and key obstructions

☐ Fire walls and blank walls

☐ Stairways (open or enclosed)

☐ Elevators (open or enclosed)

☐ Furnaces or boilers

☐ Exterior ornamentation such as marquees, awnings, and billboards

☐ Fire escapes

☐ Chimneys or stacks

☐ Water supplies (size, capacity, source, pump, tank)

☐ Underground fuel tanks and power lines

☐ Overhead power lines and obstructions

☐ Areas protected by sprinklers

☐ FDC or wall hydrants

☐ Risers (with sizes)

☐ Standpipes (with sizes)

☐ Check valves

☐ Gate valves

☐ Floor construction

☐ Significant under-floor conditions

☐ Roof type, support, and composition

☐ Building construction and any building modifications

☐ Significant ground slope

☐ Title block

☐ Sectional drawing if appropriate

☐ Legend

Information Sheet 19-4
Field Sketch Work Sheets

1 inch equals ___ feet   Scale: _____

(50 millimeters equals ___ meters)

Date _____   Initials _____

◯ North (insert arrow)

Reference:

Location:

Permission is granted to duplicate this sketch sheet.

1 inch equals ___ feet   Scale: _____

(50 millimeters equals ___ meters)

Date _____   Initials _____

◯ North (insert arrow)

Reference:

Location:

Permission is granted to duplicate this sketch sheet.

Practical Activity Sheet 19-1
Make Field Sketch and Report Drawings

Name _____ Date _____

Evaluator _____ Overall Competency Rating _____

References NFPA 1001, Prevention, Preparedness, and Maintenance 4-5.1
Essentials, pages 659–666.

Prerequisites None

Introduction As a member of a pre-incident survey team, you should know report drawing techniques and standard plan symbols and abbreviations. With this knowledge, you will be able to develop accurate field sketches and report drawings as assigned.

Directions Draw field sketch and report drawings.

Activity
1. Gather the necessary equipment for developing field sketch and report drawings.
   - Drawing pencils
   - Drawing board
   - Graph paper
   - Architect's scale
   - Protractor
   - Straight edge rule
   - *NFPA Handbook* and department's mapping symbols
   - Quiet place to work

2. Study the field sketch and report drawings in Information Sheets 19-1 and 19-2.

3. Choose a community building to which you have access — public library, school, hospital, store, office building, business.

4. Measure and make a floor plan, plot plan, or elevation field sketch (or all three) of the chosen floor or building. Use the checklist and the building plans sketch sheet in Information Sheets 19-3 and 19-4. You may duplicate the sketch sheets as necessary.

5. Return to the station and make a report drawing of the field sketch. Use the checklist in Information Sheet 19-3. Be sure to use standard mapping symbols and abbreviations and draw the floor plan to scale. Include a key or legend of the symbols used.

Competency Rating Scale

3 — Skilled — "Yes" checked for all criteria; student requires no additional practice.

2 — Moderately skilled — "Yes" checked for all critical (starred) criteria; student may benefit from additional practice.

1 — Unskilled — "No" checked for one or more critical (starred) criteria; student requires additional practice and reevaluation.

☒ — Unassigned — Task is not required or has not been performed.

✔ **Evaluator's Note:** Score the product as indicated below. Use the rating scale above to assign an overall competency rating. Record the overall competency rating on both the student's practical activity sheet and competency profile.

To show competency in this objective, the studnt must achieve an overall rating of at least 2.

Criteria	Yes	No
Field Sketch		
*1. Legible	☐	☐
*2. Measurements accurate	☐	☐
*3. Basic information complete	☐	☐
Report Drawing		
1. Good overall appearance	☐	☐
*2. Accurate	☐	☐
*3. Scale correct and appropriate	☐	☐
*4. Use of standard symbols/abbreviations correct	☐	☐
*5. Complete and adequately detailed	☐	☐
6. Legend useful	☐	☐
*7. Complies with SOPs	☐	☐

* Critical criterion

Job Sheet 19-1
Perform a Pre-Incident Survey and Complete Related Documentation

Name _____ Date _____

Evaluator _____ Overall Competency Rating _____

References | NFPA 1001, Prevention, Preparedness, and Maintenance 3-5.1a
Essentials, pages 659–666

Prerequisites
- Chapter 3, Building Construction
- Practical Activity Sheet 19-1 — Make Field Sketch and Report Drawings
- Job Sheet 15-1 — Inspect Protected Property Fire Suppression Systems
- Working knowledge of your jurisdiction's codes and standards

Student's Instructions | To meet evaluation standards, you must perform this job within _____ *[amount of time, if applicable];* you may have _____ attempts. When you are ready to perform this job, ask your instructor to observe the procedure and complete this form. To show mastery of this job, you must perform all steps to receive an overall competency rating of at least 2.

Competency Rating Scale

3 — Skilled — Meets all evaluation criteria and standards; performs task independently on first attempt; requires no additional practice or training.

2 — Moderately skilled — Meets all evaluation criteria and standards; performs task independently; additional practice is recommended.

1 — Unskilled — Is unable to perform the task; additional training required.

⊠ — Unassigned — Job sheet task is not required or has not been performed.

✔ **Evaluator's Note:** Formulate and inform the candidate of the standards for this task (time allowed and number of attempts). Observe the candidate perform the task, check the step/key point under the appropriate attempt number as accomplished, record total time (if appropriate), and then use the rating scale above to assign an overall competency rating. If the candidate is unable to perform any step of this job, have the candidate review the materials and try again.

Introduction | Pre-incident survey procedures are usually considered the most important activity — aside from fire fighting — performed by firefighters. A carefully planned survey program carried out by well-trained personnel can reduce the loss of life and property should an emergency occur.

Equipment and Personnel
- Neatly groomed and uniformed survey team
- Coveralls for crawling into attics and confined spaces
- Hard hat
- Steel-toed shoes
- Eye protection
- Gloves

- Any other items appropriate for personal protection
- Clipboard, inspection forms, and standard plan symbols
- Pencils and paper (Information Sheet 19-4) for preparing sketches
- 50-foot *(15 m)* tape measure
- Flashlight (most concealed spaces are not lighted)
- Camera equipped with flash attachment and loaded with film
- Pitot tubes and gauges for water flow tests
- Copy of fire code and inspection manuals

		Attempt No.		
Job Steps	**Key Points**	**1**	**2**	**3**
1. Enter the premises.	1. a. At the main entrance	—	—	—
	b. When summoned by company officer	—	—	—
2. Return to the outside of the building.	2. a. After introductions	—	—	—
	b. After permission for survey has been obtained	—	—	—
	c. At direction of your company officer	—	—	—
3. Record in preliminary notes your initial observations before entering premises.	3. a. Number and location of fire hydrants, fire alarm boxes, and standpipes and connections	—	—	—
	b. Type of building construction and materials	—	—	—
	c. Height, occupancy, construction, and materials of adjacent exposures	—	—	—
	d. General housekeeping of area around occupancy	—	—	—
	e. Condition of the access streets	—	—	—
	f. General appearance of neighborhood	—	—	—
	g. Type of building and occupancy	—	—	—
	h. Address numbers for visibility	—	—	—
	i. All sides of building for accessibility	—	—	—
	j. Threats from natural encroachments	—	—	—

Job Steps	Key Points	Attempt No. 1	2	3
	k. Forcible entry problems posed by barred windows or high-security doors	—	—	—
	l. Overhead obstructions that would restrict emergency operations	—	—	—
4. Draw exterior field sketches (or revise existing sketches) to include all pertinent data.	4. Per Practical Activity Sheet 19-1	—	—	—
5. Calculate and record hydrant flow rates as appropriate.	5. Using pitot tube and gauges	—	—	—
6. Take photographs.	6. a. With owner's or representative's permission	—	—	—
	b. Shooting building from above, when possible, to provide an overall view	—	—	—
7. Survey the building interior.	7. a. Beginning either on lowest floor or roof	—	—	—
	b. Surveying each room on each floor in succession	—	—	—
8. Take notes while surveying premises.	8. a. General housekeeping	—	—	—
	b. Employee education in location and use of fire prevention equipment and evacuation procedures	—	—	—
	c. Life safety information (hazardous practices or substances)	—	—	—
	d. Condition and maintenance dates of fire protection systems (standpipes, sprinkler systems, alarms, fire extinguishers)	—	—	—
9. Draw floor plan field sketches (or revise existing floor plans) to include all pertinent data.	9. a. For each floor	—	—	—
	b. Per Practical Activity Sheet 19-1	—	—	—
	c. To include the following: — Location of permanent walls, partitions, fixtures, machinery (do not include furniture)	—	—	—

Essentials (FF II)   **Job Sheet**   **SA 19 - 433**

Job Steps	Key Points	Attempt No. 1	2	3
	— Locations of vertical shafts or horizontal openings (stairways, escalators, elevators, ventilation ducts, laundry chutes, incinerator shafts)	—	—	—
	— Location, sizes, types of fire protection equipment (standpipes, sprinkler systems, fire extinguishers)	—	—	—
	— Stored flammable substances and toxic interior finishes	—	—	—
10. Draw detailed sketches.	10. As necessary to show complex construction	—	—	—
11. Take photographs.	11. a. With owner's or representative's permission	—	—	—
	b. Close-ups to show detailed features	—	—	—
12. Survey, take notes, and sketch all other structures on the property.	12. In same manner as for primary structure	—	—	—
13. Conduct an exit interview.	13. a. Obtaining all necessary tenant and property conservation information:			
	— Structure address	—	—	—
	— Street location	—	—	—
	— Type of business and content hazards	—	—	—
	— Owner/tenant information	—	—	—
	— Whom to contact on premises	—	—	—
	— The working hours	—	—	—
	— The number of people who may be in the building during different times	—	—	—

Job Steps	Key Points	1	2	3
	— Special rescue problems such as handicapped, ill, very old, or very young occupants	—	—	—
	— Contents with a high value	—	—	—
	— Items that could be harmed by water	—	—	—
	— Facilities that could be used to aid in salvage operations	—	—	—
	b. Commenting favorably on the good conditions found	—	—	—
	c. Thanking the occupant for the courtesies extended to the fire department	—	—	—
	d. Answering any questions that the occupant may have	—	—	—
	Time (Total)	—	—	—

Evaluator's Comments _____

_____

_____

Chapter 19 Review Test

> ⇢ **Directions:** This review test covers the Firefighter II material in Chapter 19 of your ***Essentials of Fire Fighting*** text. It may be assigned as a study aid (self-test) or may be administered by your instructor as a pretest or posttest.
>
> When used as a study aid, try to answer the questions without referring to the page numbers in ***Essentials*** or your ***Firefighter II Student Applications*** workbook *(SA)* on which the answers can be found until after you have completed the entire test. Then check your answers against those on the pages provided in parentheses.
>
> When administered by your instructor as a pretest or posttest, read each of the test questions carefully. Choose the best response and then darken the corresponding letter on your answer sheet.
>
> This chapter review test contains 30 multiple-choice questions, each worth 3 points. To pass the test, you must achieve at least 75 of the 90 points possible.

1. Which of following pieces of equipment is ***not*** used during a survey? *(660, 661)*

   A. 50-foot *(15 m)* tape measure

   B. Current meter with surge detection capability

   C. Pitot tube and gauges

   D. Camera equipped with flash attachment

2. How should building surveys be scheduled? *(661)*

   A. With no advance notice in order to catch the facility under normal occupancy loads

   B. During a period of the day or evening when the building is empty

   C. With advance notice during the evening or after working hours for the owner

   D. With advance notice during a period of the day or evening when the building is under its normal occupancy load

3. Which of the following statements is false with regard to the exterior portion of the survey? *(662, 663)*

   A. The firefighter should look for the location of fire hydrants.

   B. The firefighter should determine drainage system flow rates.

   C. The firefighter should look for the location of fire alarm boxes.

   D. The firefighter should determine the location and number of possible exposures.

4. What should a firefighter do first after receiving permission from the owner to survey the property? *(663)*

   A. Conduct a survey of the building interior.

   B. Check the condition and operation of all private fire protection systems.

   C. Proceed directly to a roof or basement to begin a systematic survey.

   D. Conduct a survey of the building exterior.

---

5. Firefighter A says that a survey team should check for overhead obstructions that would restrict aerial ladder operation.

Firefighter B says that a survey team should check the distance to the nearest water storage tank before entering the property.

Who is right? *(662, 663)*

A. Firefighter A
B. Firefighter B
C. Both A and B
D. Neither A nor B

6. What should a survey team do if refused entry to any area? *(663, 664)*

A. Record the name of the person refusing entry, and explain that the fire department will not provide fire protection to those buildings denied open survey.

B. Report the person's name and address to the station safety officer.

C. Suspend the survey, record the name and address of the person refusing entry, and report the incident to the police.

D. Report the incident to the fire chief or fire marshal for appropriate action.

7. What should the firefighter do if a floor plan was made during a previous building survey? *(664)*

A. Disregard the old floor plan.

B. Obtain a new floor plan from the occupant or building owner.

C. Modify the existing floor plan to reflect changes.

D. Use the existing floor plan as it is.

8. Which of the following statements is false? *(661, 662, 664)*

A. Firefighters should ignore hazardous materials during surveys because a separate hazardous materials fire inspection is required for facilities that house such substances.

B. The company officer should contact the building occupant prior to the survey to set up a suitable day and time for the survey.

C. When making a survey, firefighters should enter the premises by the main entrance.

D. If the property includes several buildings, each should be surveyed separately.

9. In what sequence should a survey of a building's interior be performed? *(663)*

A. Roof downward
B. Basement upward
C. Either A or B
D. Perimeter to center

10. Which of the following statements best explains why a survey team should meet with the person in authority before leaving the premises after a pre-incident survey? *(666)*

A. To ensure that the person in authority immediately corrects any areas that are in noncompliance with regulations

B. To allow the person in authority to defend or justify any areas of noncompliance

C. To maintain a cooperative relationship with the person in authority

D. To allow the person in authority to provide a written evaluation of the survey team

11. Firefighter A says that the survey team should point out any negative findings during the exit interview and follow up with a formal written citation.

Firefighter B says that if the survey team cannot answer an owner/occupant's questions, they should refer the owner/occupant to the fire marshal.

Who is right? *(666)*

A. Firefighter A   C. Both A and B

B. Firefighter B   D. Neither A nor B

12. What does the standard map symbol below portray? *(665)*

A. An automatic fire alarm

B. An infrared sprinkler system

C. A fire alarm gong

D. A fire alarm box

(FA)

13. Firefighter A says that on-site sketches should be drawn to scale on graph paper using rulers, squares, triangles, and a portable drawing table if available.

Firefighter B says that by using standard map symbols on a sketch, the firefighter can show the type of construction, thickness of walls, partitions, openings, roof types, parapets, and other important features.

Who is right? *(664)*

A. Firefighter A   C. Both A and B

B. Firefighter B   D. Neither A nor B

14. From a fire fighting standpoint, what photographic view of a building is especially good for survey reports? *(666)*

A. Elevated

B. Overhead

C. Wide-angle

D. Close-up

15. Which of the following is *not* true with regard to making plot plans? *(664)*

A. A plot plan shows the general arrangement of the property with respect to streets, other buildings, and other features.

B. Plot plans need not be neat or accurate because they are only sketches.

C. Plot plans should be made for surveyed properties for which there are no existing accurate maps.

D. Plot plans often constitute the most informative part of a survey.

16. To leave the premises without consulting the person in authority ___. *(666)*

A. Is a compliment that acknowledges how important that person's time is

B. May give the impression that the survey was not important

C. Means that the formal report will have to be submitted to the owner/occupant by mail

D. Is recommended if the survey results are negative

17. Which of the following statements is **not** true about sectional elevation sketches? *(665, 666)*

   A. Sectional elevation sketches may be needed to show elevation changes, mezzanines, balconies, and other features.

   B. If a sectional elevation sketch along an exterior wall does not show important features, the sketch need not be done because any other view would be too confusing.

   C. The easiest sectional view to portray is to establish the imaginary line along an exterior wall.

   D. The permanent sectional elevation drawing should be drawn to scale.

18. Photographs can be useful to surveys when used to show ___. *(666)*

   A. Exterior views

   B. Interior views and close-ups

   C. Both A and B

   D. Neither A nor B; photographs should not be used for formal surveys

19. What does the standard map symbol below indicate? *(665)*

   A. Partition wall

   B. Public water service

   C. Private water service

   D. Concrete or masonry wall

   6" W.P. (PRIV.)

20. What does the standard map symbol below indicate? *(665)*

   A. Iron chimney

   B. Fire pump

   C. Vertical steam boiler

   D. Fire department connection

21. What does the standard map symbol below indicate? *(665)*

   A. Open hoist

   B. Skylight

   C. Stairwell

   D. Fire detection system

22. Firefighter A says that the success of pre-incident planning depends on the firefighter's technical ability to perform adequate building surveys.

   Firefighter B says that the success of pre-incident planning depends on the firefighter's ability to work well with building owners and occupants.

   Who is right? *(660)*

   A. Firefighter A

   B. Firefighter B

   C. Both A and B

   D. Neither A nor B

23. What type of plan is illustrated below? *(665)*

    A. Plot

    B. Detail

    C. Floor

    D. Elevation

24. What does the standard map symbol below indicate? *(665)*

    A. Standpipe

    B. Gasoline tank

    C. Single hydrant

    D. Sprinkler riser

25. Firefighter A says that firefighters must obtain permission from the owner/occupant to conduct a building survey.

    Firefighter B says that a representative of the occupancy should not be allowed to accompany firefighters during a building survey.

    Who is right? *(661, 662)*

    A. Firefighter A   C. Both A and B

    B. Firefighter B   D. Neither A nor B

26. Which of the following items should a team take when water flow tests will be required during a survey to check a water-based fire protection system? *(661)*

    A. Flashlights   C. Hoses and buckets

    B. Pitot tubes and gauges   D. Sprinkler kit

27. Which of the following is *not* a method by which firefighters can improve their survey skills? *(660)*

    A. Improved public speaking through good grooming and personal appearance

    B. Improved ability to judge conditions through study, experience, and on-the-job training

    C. Improved ability to transpose visual information into written reports with time and practice

    D. Improved insight into difficult situations through counseling by company officers and fire prevention officer

28. Which of the following is *not* a guideline for scheduling a fire safety survey? *(661)*

    A. Contact the owner or occupant ahead of time to arrange the survey.

    B. The company officer should inform the owner/occupant of the purpose of the survey.

    C. Schedule the survey during nonbusiness hours to avoid congestion.

    D. The fire department administration should set a schedule for survey activities.

29. When should the survey team be introduced to the owner/occupant? *(662)*

    A. After completing the exterior survey and upon entering the building

    B. During the exit interview

    C. Upon arriving at the site

    D. None of the above; the owner/occupant should not be bothered with introductions of team members

30. How should a survey team respond if told that a restricted area contains confidential or secure processes? *(663, 664)*

    A. State that the occupant must allow admittance under fire regulations.

    B. Terminate the survey and report the situation to the fire marshal.

    C. Make a note in the survey report to indicate that the details of that area are unknown.

    D. Suggest that the process be covered or screened to permit the survey to continue.

No part of this test may be reproduced without written permission from the publisher.

REVIEW TEST ANSWER SHEET

	A	B	C	D
1.	○	○	○	○
2.	○	○	○	○
3.	○	○	○	○
4.	○	○	○	○
5.	○	○	○	○
6.	○	○	○	○
7.	○	○	○	○
8.	○	○	○	○
9.	○	○	○	○
10.	○	○	○	○
11.	○	○	○	○
12.	○	○	○	○
13.	○	○	○	○
14.	○	○	○	○
15.	○	○	○	○
16.	○	○	○	○
17.	○	○	○	○
18.	○	○	○	○
19.	○	○	○	○
20.	○	○	○	○
21.	○	○	○	○
22.	○	○	○	○
23.	○	○	○	○
24.	○	○	○	○
25.	○	○	○	○
26.	○	○	○	○
27.	○	○	○	○
28.	○	○	○	○
29.	○	○	○	○
30.	○	○	○	○
31.	○	○	○	○
32.	○	○	○	○
33.	○	○	○	○

	A	B	C	D
34.	○	○	○	○
35.	○	○	○	○
36.	○	○	○	○
37.	○	○	○	○
38.	○	○	○	○
39.	○	○	○	○
40.	○	○	○	○
41.	○	○	○	○
42.	○	○	○	○
43.	○	○	○	○
44.	○	○	○	○
45.	○	○	○	○
46.	○	○	○	○
47.	○	○	○	○
48.	○	○	○	○
49.	○	○	○	○
50.	○	○	○	○
51.	○	○	○	○
52.	○	○	○	○
53.	○	○	○	○
54.	○	○	○	○
55.	○	○	○	○
56.	○	○	○	○
57.	○	○	○	○
58.	○	○	○	○
59.	○	○	○	○
60.	○	○	○	○
61.	○	○	○	○
62.	○	○	○	○
63.	○	○	○	○
64.	○	○	○	○
65.	○	○	○	○
66.	○	○	○	○
67.	○	○	○	○

	A	B	C	D
68.	○	○	○	○
69.	○	○	○	○
70.	○	○	○	○
71.	○	○	○	○
72.	○	○	○	○
73.	○	○	○	○
74.	○	○	○	○
75.	○	○	○	○
76.	○	○	○	○
77.	○	○	○	○
78.	○	○	○	○
79.	○	○	○	○
80.	○	○	○	○
81.	○	○	○	○
82.	○	○	○	○
83.	○	○	○	○
84.	○	○	○	○
85.	○	○	○	○
86.	○	○	○	○
87.	○	○	○	○
88.	○	○	○	○
89.	○	○	○	○
90.	○	○	○	○
91.	○	○	○	○
92.	○	○	○	○
93.	○	○	○	○
94.	○	○	○	○
95.	○	○	○	○
96.	○	○	○	○
97.	○	○	○	○
98.	○	○	○	○
99.	○	○	○	○
100.	○	○	○	○

Name _____

Date _____

Score _____

Chapter 19 Competency Profile

Student Name _____ Soc. Sec. No. _____

_{Last   First   Middle}

Fire Department _____

Address _____

Phone _____

Home Address _____

Phone _____

Date of Enrollment _____ - _____ - _____   Total Class Hours _____

Date of Withdrawal _____ - _____ - _____   Total Hours Absent_____

Date of Completion _____ - _____ - _____

Instructor's Name _____ Session Dates_____

Instructor's Directions

1. Check the student's competency rating (3, 2, 1, ☒) for each performance test task and psychomotor lesson objective (practical activity and job sheets) listed below.

2. List any additional performance tasks or psychomotor objectives (job sheets or practical activity sheets) under "Other," and check the candidate's competency rating.

3. Record the candidate's cognitive scores (written lesson tests and *administered* chapter review tests) in the spaces provided.

Level				Psychomotor Competencies
3	2	1	☒	

Practical Activity Sheet

3	2	1	☒	
☐	☐	☐	☐	PAS 19-1 — Make Field Sketch and Report Drawings
☐	☐	☐	☐	Other _____
☐	☐	☐	☐	_____

Job Sheet

3	2	1	☒	
☐	☐	☐	☐	JS 19-1 — Perform a Pre-Incident Survey and Complete Related Documentation
☐	☐	☐	☐	Other _____
☐	☐	☐	☐	_____

Level			
3	2	1	☒
☐	☐	☐	☐
☐	☐	☐	☐
☐	☐	☐	☐
☐	☐	☐	☐

Psychomotor Competencies

Chapter 19 Performance Test

Task 1 — Working as a member of a team, conduct a pre-incident survey.

Task 2 — Update pre-incident planning files.

Other _____

_____

Points Achieved	Points Needed/ Total

Cognitive Competencies

Written Test

_____ 5/6	1. Provide examples of personal traits and skills required of personnel who conduct fire safety surveys.
_____ 8/10	2. Provide examples of the type of equipment required to conduct fire safety surveys.
_____ 5/5	3. List goals of pre-incident surveys.
_____ 12/15	4. Provide examples of the types of information that a pre-incident survey can provide.
_____ 8/10	5. Match standard map symbols to their correct meanings.
	6. Evaluated on Practical Activity Sheet 19-1
_____ 5/5	7. List objectives of the exit interview during a pre-incident survey.
	8. Evaluated on Job Sheet 19-1

Review Test

_____ Chapter 19 Review Test

Instructor's Signature _____ **Date** _____

Student's Signature _____ **Date** _____